Inside MATHEMATICS

Algebra to Calculus

UNLOCKING MATH'S AMAZING POWER

$$x = \frac{-b \pm \sqrt{b^2 - 4ac}}{2a}$$

Inside **MATHEMATICS**

Algebra to Calculus

UNLOCKING MATH'S AMAZING POWER

Mike Goldsmith
SERIES EDITOR: TOM JACKSON

Shelter Harbor Press
NEW YORK

Introduction	6
The Dawn of Algebra	10
Proof	16
The Pythagoreans	26
The Algebra of Shapes	34
The Prehistory of Calculus	40
Equations	46
The Third Dimension	54
Algebra Moves East	60
Cubics	64
Sequences and Series	66
Unreal Numbers	74
The Rules of Algebra	82
Finding the Maximum	86
Algebraic Geometry	92
Fermat's Last Theorem	98

Pascal's Triangle	102
Calculus	110
Differential Equations	116
e	122
The Fundamental Theorem of Algebra	130
The Fundamental Theorem of Calculus	136
Groups	140
Quaternions	148
The Mathematics of Thought	152
Abstract Algebra	158
Paradoxes of Zeno, Russell, and Gödel	166
7 Millennium Problems	174
Calculus in Depth	176
Glossary	180
Index	182
Credits	184

Introduction

It is probably safe to say that algebra and calculus are two of the least popular parts of math, and many of us have gazed with anxiety at the letters and symbols revealed as we flicked through the pages of school math books for the first time. The two very good reasons for this dismay could be summed up in the two questions that most new students ask:

1. What does it all mean?
The main thing that makes algebra and calculus so off-putting is that neither is written in plain English. Or at least, the important parts aren't. Unlike arithmetic, which is concerned with familiar numbers, or geometry, where diagrams give plenty of clues as to what is going on, algebra and calculus are very largely composed of arrangements of letters and symbols used in ways we rarely come across in normal life.

2. What's the point?
Gripping titles like "Become a Calculus Millionaire" or "Explore the Solar System through Algebra" are rarely found on bookshelves, and reading through the contents pages of algebra or calculus books does not usually help the reader see the usefulness of the topics covered. Unlike chapter titles in a novel, they are not likely to encourage anyone to explore further. *Contrapositive*, *factorization*, *integration*, and *differential equations* are not the kind of words to start many of us turning excitedly to the next page.

Who's to blame for algebra? This guy, Al-Khwarizmi, who first used the word in 830 CE. But really, we should thank him for it. Read on to find out why.

Answers

The job of this book is to answer both those questions fully, but we can give partial answers straightaway.

In a way, algebra is a language, but it differs from a language like English because it has been developed over many centuries for a very particular purpose: the explanation, analysis, and solution of the most challenging problems in many areas of life, including engineering, physics, and economics. We can discuss such problems in English too of course, but mathematical methods are more precise, and the language of algebra is also more exact than English. Calculus uses the same language as algebra does.

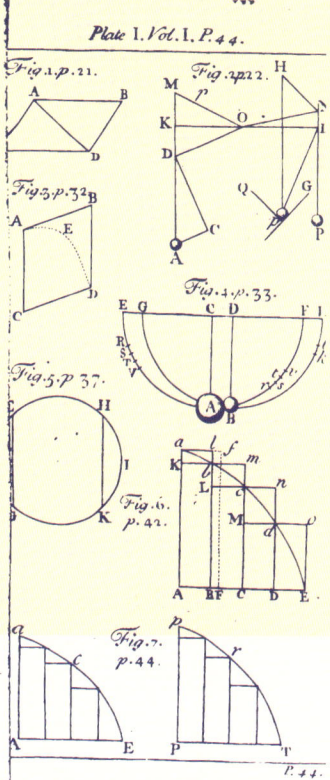

Algebra and calculus allow us to understand how nature changes.

We might want to answer the question "I want to go for a swim at the new pool (which is 1 mile away) but I only have 2 hours before I need to be back here again. Is it worth going?" Algebraically, we can capture the essentials of the situation as an equation:

$$t_{travel} + t_{change} + t_{swim} + t_{dry\ and\ change} + t_{travel\ home} = 2\ hrs$$

Faced with any puzzling equation, the first thing to do is to try to simplify it. Assuming it takes the same time to go to the pool as to come back ($t_{travel} = t_{travel\ home}$), and the same time to get changed before and after the swim if you hurry ($t_{change} = t_{dry\ and\ change}$), we can simplify and say:

$$2t_{travel} + 2t_{change} + t_{swim} = 2\ hrs$$

Let's take t_{travel} first. Since people walk at about 3 miles per hour, and the pool is 1 mile away, we can work out how long t_{travel} is. How? Well, the further one walks the longer a walk takes. Algebraically we can write this as **travel time ∝ travel distance**, which means "travel time is proportional to distance traveled." To make this more concise, we can abbreviate it to **t ∝ d**. But of course it's not just distance that we need to consider, but speed, too. The faster we walk,

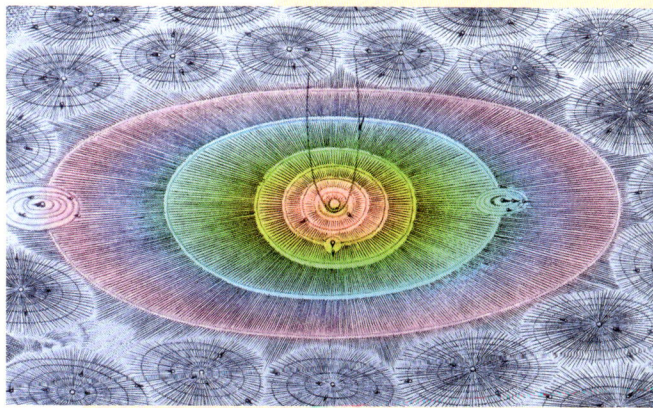

Advanced math creates imaginary realms that still tell us about the here and now.

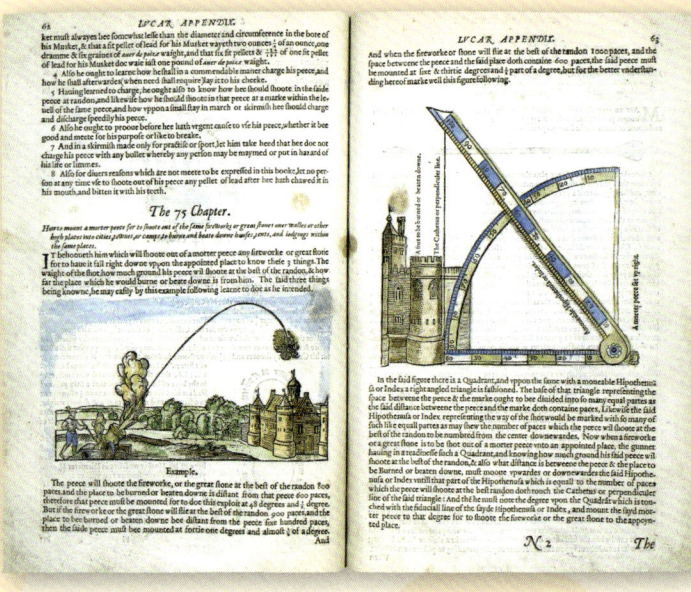

The roots of algebra and all math lie in finding solutions to everyday problems.

in the formula in place of the **d** and the **s**. Here **d** and **s** are called variables, as they can be replaced with numbers, the choice of which varies with the particular problem.

t_{travel} = ⅓ hour

and, if it takes quarter of an hour to get changed, then t_{change} = ¼ hour.

the shorter the time that is needed for traveling. This is written as **t** ∝ **¹/s**, which is "travel time is inversely proportional to travel speed."

To see why it's written as a fraction, consider these fractions: ½, ⅓, ¼, ⅕, ⅙. They get smaller from left to right, because the number under the line gets bigger. That is to say, the size of the fraction is inversely proportional to the number under the line.

So, we have:
t ∝ **d** and **t** ∝ **¹/s**

As the time for the journey is related just to these two things, we can put them together into a complete formula:

t = ᵈ/s
As we already know that **d = 1** mile and **s = 3** miles per hour, we can insert these values

We just put these values into our earlier equation: $2t_{travel}$ + $2t_{change}$ + t_{swim} = **2** hours

²/₃ + ²/₄ + t_{swim} = **2** hours

²/₃ of an hour is **40** minutes, and ²/₄ of an hour is the same as ½ hour, which is **30** minutes. As we are now talking in minutes, we will change the **2** hours into minutes too, to give us
40 minutes + **30** minutes + t_{swim} = **120** minutes.
Which means **70** minutes + t_{swim} = **120** minutes.
It's fairly easy to see from this that t_{swim} must be **50** minutes; this is obtained by subtracting **70** minutes from both sides: **70** minutes − **70** minutes + t_{swim} = **120** minutes − **70** minutes.

Therefore t_{swim} = **50** minutes.

So that's algebra. A useful tool for problems of all kinds, everyday specifics like the above, and more general ones like the amount that time

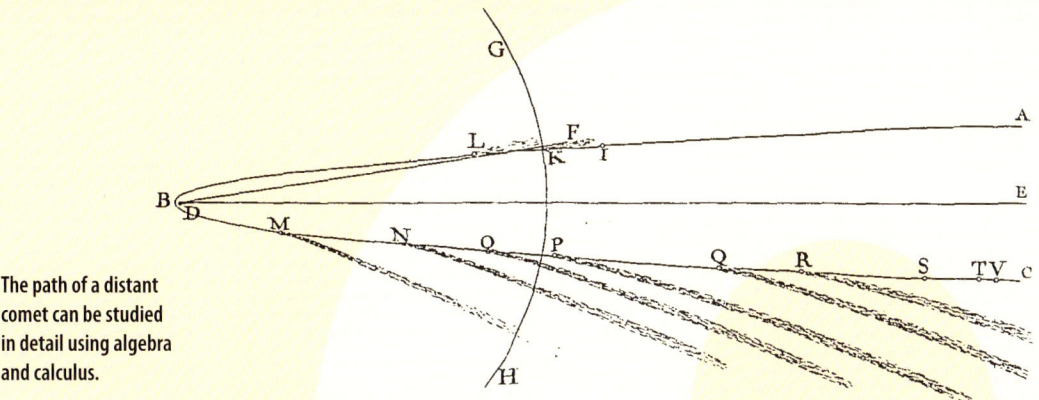

The path of a distant comet can be studied in detail using algebra and calculus.

slows down near a massive object like a star:

$$t_0 = t_f \sqrt{1 - \frac{2GM}{rc^2}}$$

(t_0 is time as measured near the object, t_f is the time as measured far from it, G is a constant, M the mass of the object, r the distance from the object at which t_0 is measured and c is the speed of light). A constant is the same number in every calculation.

And calculus?

Calculus is the key mathematical way for scientists, engineers, and economists to get to grips with the world.

In the above example we worked out how long it took to walk somewhere at a given speed. Speed is a rate of change of position—and change is what nature is all about. The economy, planets, cars, the world's population … all change with time, and to get a handle on them we need the right mathematical tool. And that tool is calculus.

One of the inventors of calculus was Isaac Newton, and he invented it as just that—a tool to tell him precisely how planets and comets changed their positions, speeds, and paths under the influence of the force of gravity.

But algebra and calculus aren't just powerful tools to solve practical problems. For reasons which no one understands, our cosmos behaves according to the same rules that govern mathematics, and many concepts invented by mathematicians for their own use have turned out to correspond to reality (such as imaginary numbers, see page 79). So mathematics really does offer us the keys to understanding the universe.

The math on this postcard (written by Emmy Noether) shows how space and time are linked together.

The Dawn of Algebra

IN ANCIENT BABYLONIA (IN WHAT IS NOW IRAQ) ABOUT 4,000 YEARS AGO, ALGEBRA BEGAN. We know this because of the Babylonians' love of record-keeping and also because they kept those texts in the form of rock-hard clay tablets and cylinders that have survived the ages.

Babylonian writing was made by pressing a specially shaped stylus into wet clay, leaving wedge-shaped patterns. The writing system they used is now called cuneiform, and it remained popular in many civilizations for more than a thousand years. The Babylonians were such dedicated writers that over half a million of their clay tablets survive. By the 1860s, it was clear that many contained number symbols, but these were not regarded with much interest at the time.

Revealing the past

We know of no individual Babylonian mathematicians, but there is a person whose name is inextricably linked with cuneiform algebra—the Austrian mathematician Otto Neugebauer. It was he who untangled the

Babylon is famed for its Hanging Gardens, now long turned to dust. Its math has lasted much longer.

calculations in the clay and pieced together an understanding of Babylonian algebra, publishing his findings in the 1930s and 1940s. Neugebauer worked in Germany, where his work was highly respected, and was appointed head of the illustrious Göttingen Mathematical Institute in 1933. However, on his very first day there he was asked to sign an oath of loyalty to the government, which had just come under the

BABYLONIAN NUMBERS

We use a base 10 system, which means that a four-figure number like 2,074 represents 2 thousands, no hundreds, 7 tens and 4 ones. To count in base 10, we can only use 10 different characters (including zero) before we run out: 0, 1, 2, 3, 4, 5, 6, 7, 8, 9.

At this point, we write a 1 on the left, and run through our ten characters again 10, 11, 12 … 19

Until we run out, when we increase the number on the left and start running through the ten characters once more 20, 21 …

We presumably use base 10 because we have 10 fingers and thumbs, so, if we use them to count with, we will have to come up with something else when we get past 10.

The Babylonians, however, didn't stop at 10. They used base 60, although they did not have a zero in their system.

Their system looked like this:

𒁹	1	𒌋𒁹	11	𒎙𒁹	21	𒌍𒁹	31	𒐏𒁹	41	𒐐𒁹	51
𒈫	2	𒌋𒈫	12	𒎙𒈫	22	𒌍𒈫	32	𒐏𒈫	42	𒐐𒈫	52
𒐈	3	𒌋𒐈	13	𒎙𒐈	23	𒌍𒐈	33	𒐏𒐈	43	𒐐𒐈	53
𒐉	4	𒌋𒐉	14	𒎙𒐉	24	𒌍𒐉	34	𒐏𒐉	44	𒐐𒐉	54
𒐊	5	𒌋𒐊	15	𒎙𒐊	25	𒌍𒐊	35	𒐏𒐊	45	𒐐𒐊	55
𒐋	6	𒌋𒐋	16	𒎙𒐋	26	𒌍𒐋	36	𒐏𒐋	46	𒐐𒐋	56
𒑂	7	𒌋𒑂	17	𒎙𒑂	27	𒌍𒑂	37	𒐏𒑂	47	𒐐𒑂	57
𒑄	8	𒌋𒑄	18	𒎙𒑄	28	𒌍𒑄	38	𒐏𒑄	48	𒐐𒑄	58
𒑆	9	𒌋𒑆	19	𒎙𒑆	29	𒌍𒑆	39	𒐏𒑆	49	𒐐𒑆	59
𒌋	10	𒎙	20	𒌍	30	𒐏	40	𒐐	50		

It is because our system has been developed from the Babylonian one (among others) that we still divide each hour into 60 minutes and then each minute into 60 seconds.

control of the Nazi party. He resigned at once, and left the country, working on Babylonian algebra first in Denmark and then in the USA.

Ancient math

The algebra that Neugebauer unearthed was startlingly modern in some ways. The Babylonian mathematicians were well aware of the results of Pythagoras's theorem (see page 26), and could solve quadratic equations (see box, page 13). This was despite the fact that they had no mathematical symbols of any kind, not even an equals sign. All their calculations were written out in words and numbers, a bit like recipes.

Left: The Tower of Babel in a Book of Hours from the 15th century. These books explained how to calculate the date of Christian festivals—and used Babylonian math to do it.

Above: This clay cuneiform tablet from Babylon contains 247 questions that require the quadratic equation to solve. Babylonian students obviously had good eyesight.

To work out a mathematical problem today, you might feed the appropriate numbers into a formula and use a calculator for any tricky operations. Things were very different in ancient Babylon. Instead of a textbook, you would reach for a stack of tablets, with a whole range of mathematical recipes on them, and find one that seemed similar to the problem you wanted to solve. You would then go through the steps described on the tablet, using your own numbers. You would be able to do any simple calculations yourself, but you could also refer to tables of squares and square roots when you needed to. You would also have a multiplication table to hand. Unlike today's schoolchildren, it's unlikely

How it works

Solving Quadratic Equations

Step 1: The first step is to rewrite the equation in standard form, which is $ax^2 + bx + c = 0$ so, an equation like $x^2 + 2x = 4 + 2x$ can be rewritten like this:

$x^2 + 2x = 4 + 2x$ Remove $2x$ from both sides

$x^2 = 4$

$x^2 - 4 = 4 - 4$ Subtract 4 from both sides

$x^2 - 4 = 0$

Step 2: Factorize:
$(x + 2)(x - 2) = 0$

Step 3: Insert a value of x which makes the first bracket equal zero:
$(-2 + 2)(-2 - 2) = 0$

That gives: $(0)(-4) = 0$, showing that $x = -2$ is one solution

Step 4: Insert a value for x which makes the second bracket equal zero:
$(2 + 2)(2 - 2) = 0$

Which gives: $(4)(0)$, showing that $x = 2$ is the other solution.

Step 5: For quadratics that are hard to factorize, use the Quadratic Formula:

$$x = \frac{-b \pm \sqrt{b^2 - 4ac}}{2a}$$

So, to solve $7x^2 + 3x - 11 = 0$

$$x = \frac{-3 \pm \sqrt{3^2 - 4 \times 7 \times -11}}{2 \times 7}$$

Which gives

$$x = \frac{-3 \pm \sqrt{317}}{14}$$

Which means
$x = 1.486$ (approximately)
or -1.057 (approximately)

---

that anyone would be expected to know their tables by heart: The Babylonians used a base-60 number system, which meant their multiplication tables had 59 rows and 59 columns!

One answer

The odd thing about Babylonian mathematics is that the tablets only show how to work things out, without giving any indication of how the recipes were reached in the first place. So the skill of the student here would be in selecting the appropriate example, and applying it to the particular problem in hand. Since the Babylonians did not have a concept of negative numbers, they assumed that multi-solution quadratic equations could have just one solution.

SOLVING PROBLEMS THE BABYLONIAN WAY

A typical Babylonian problem is: "The length exceeds the width by 10. The area is 600. What are the length and the width?" (The Babylonian base 60 numbers are written in base 10 here.)

Today, we would write this as:
$x - y = 10$ (1)
$xy = 600$ (2)

And solve it like this:

Rearrange (1) to give
$x = 10 + y$

And substitute into (2) to obtain
$(10 + y)y = 600$

Multiply out
$10y + y^2 = 600$

Rewrite as a quadratic equation in standard form
$y^2 + 10y - 600 = 0$

Solve this using the standard Quadratic Formula (see page 13) with **$a = 1, b = 10, c = -600$**

This gives
$x = (-10 \pm \sqrt{2500})/2$

Which is:
$x = -30$ or 20

Which means [from (1)]
$y = -40$ or 30

But the Babylonians did it by selecting a likely example from their set of tablets and following it through, inserting the numbers of the specific problem. They might end up with this:

What is the difference between the length and the width? **(10)**

Half this **(5)**
Square this **(25)** (The Babylonian student would find this in a table of square roots.)
Add the area **(625)**
Find the square root **(25)**
Add half of the difference between the length and the width to this square root **(30)**. This is the length.

Subtract half of the difference between the length and the width from the square root **(20)**. This is the width.

Babylonian architects and engineers relied on advanced math to create their sturdy and impressive buildings.

The shape of the Pyramids of Giza shows that ancient Egyptians had a firm grasp of math.

Secret of success

There had been advanced civilizations for many thousands of years before the Babylonians, and all were able to count. But, as far as we know, none of their expertise came close to that of the Babylonians. One thing that made mathematics far easier for the Babylonians than for any earlier civilization (and some later ones, too) is that they used the positional number system, which they seem to have invented. This is one in which the same symbol takes on different values according to where is placed. Our system is a positional one, too. The numbers 246, 426 and 642 each mean something different, despite the fact they each use the same three characters, and we can "decode" them because we know what each position means: The first position means "hundreds," the next "tens," and the third position means "ones." This makes numbers easy both to read and to manipulate. We don't have any records that show how the Babylonian mathematicians worked out their recipes and as far as we know, they had no concept of a general solution with variables. To them, Equations (1) or (2) in the box on the left would be meaningless. So, while they were familiar with Pythagorean triples, (whole numbers that obey $a^2+b^2=c^2$) such as 3, 4, 5 or 5, 12, 13, and include examples of them in their texts, the idea that they could all be represented by a single formula lay centuries ahead.

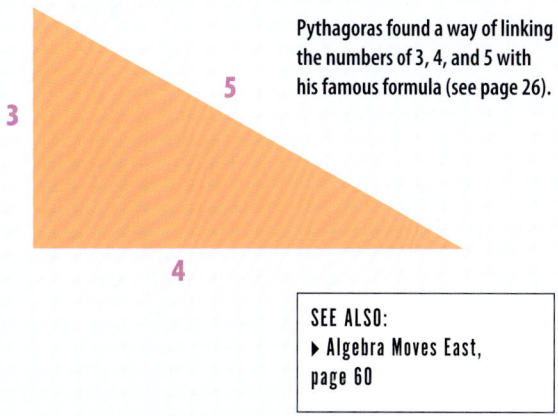

Pythagoras found a way of linking the numbers of 3, 4, and 5 with his famous formula (see page 26).

SEE ALSO:
▶ Algebra Moves East, page 60

Proof

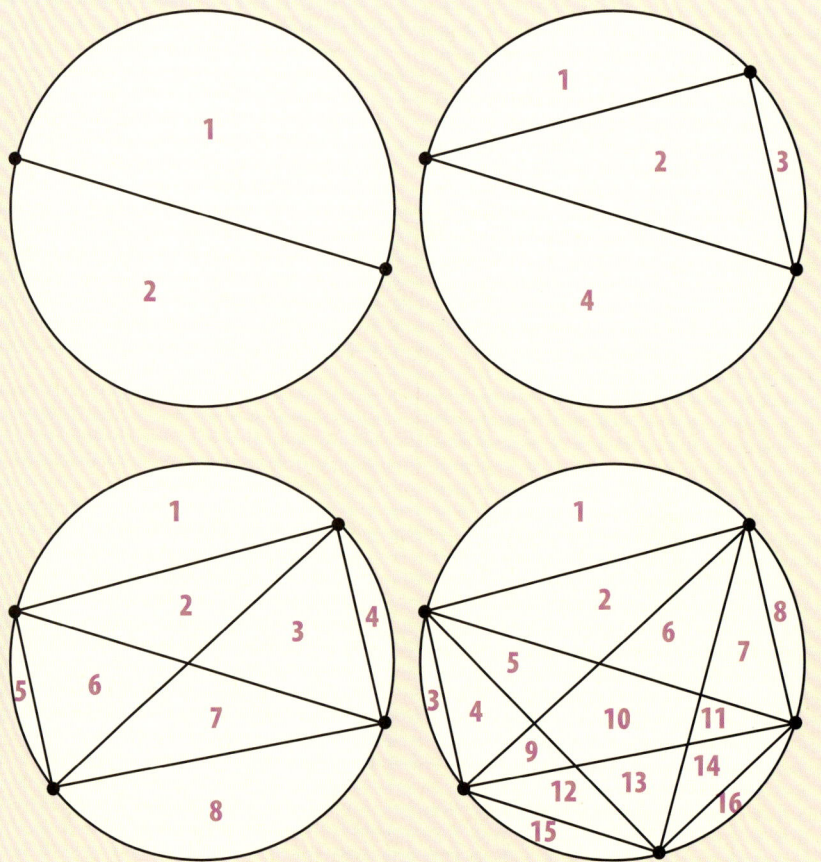

THEOREMS ARE THE RULES OF MATHEMATICS, and they range from fairly simple, like the Pythagorean theorem, to those that are so complex that no one knows whether they are true or not.

To be of use to mathematicians, a theorem has to be proved. There are five main ways to do this, which we will look at later on. Of course, Babylonian mathematicians used rules all the time, but the idea of stating these rules as theorems, or trying to prove them, would have seemed very strange to them. If you could ask a Babylonian mathematician how he or she knew the right way to carry out a calculation, the answer would probably be "because it works" or "because that's the way it's always been done." This is like the way in which most of us learn to speak our own language. We find out what works, and just get the hang of it. When we see a new word and have to pronounce it, we don't need to be told what the rule is. Take a made-up word like "zam." There's a very good chance it rhymes with "ham."

Dividing a circle seems to follow a pattern. But mathematics demands proof of what "seems" to be true.

And if you add an "e," the resulting word will probably rhyme with "claim." So, there are rules to language even though we may never spell them out to ourselves—but when we learn a new language, we do need to learn the rules for it.

Math rules

In math, rules are vital to make progress; it's not safe just to go with what seems to work. For instance, in the four circles above, each dot is connected by straight lines to all the other dots.

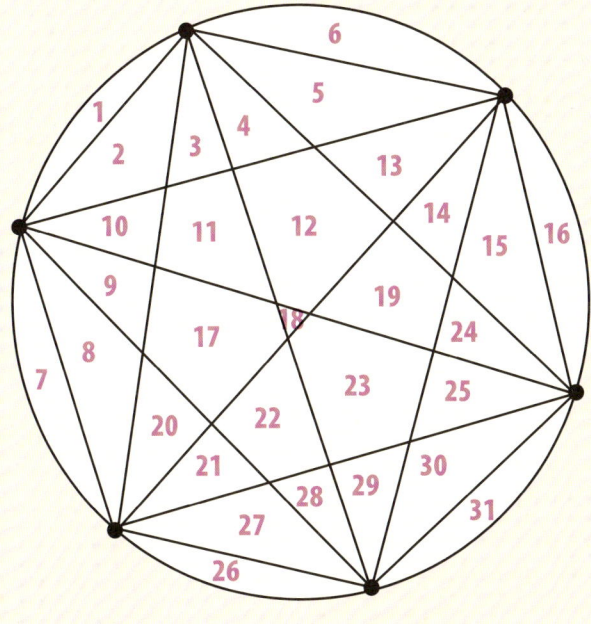

And we can check this formula:

dots	dots−1	2^(dots−1)
2	1	2
3	2	4
4	3	8
5	4	16

How many areas do you think a 6-dotted circle will contain? The formula tells us it will be 32. However, the answer is 31—take a look, left. So, this goes to show that a theorem can't be trusted just because it seems to work. It needs more than just testing—it needs to be proved, too.

Math's first star

The first mathematician in recorded history was Thales, and as far as we know, he was the first person to prove theorems (including the one

The numbers of areas into which the lines divide the circles are:

2 dots : 2 areas

3 dots : 4 areas

4 dots : 8 areas

5 dots : 16 areas

It looks like a pattern is emerging. The number of areas is doubling each time a dot is added. We can even come up with a formula to describe what is happening:

number of areas = $2^{(\text{number of dots} - 1)}$

Thales, a scientist, mathematician, and philosopher, lived in Miletus (now Turkey) about 2,600 years ago.

named after him, see box, right). Thales was a Greek who was born in Miletus, which is now part of Turkey. As with many ancient Greeks, there are plenty of stories about him, but we don't know how much truth there is in them. According to one, he fell down a well one night because he was so busy looking at the stars, and he is also said to have made a small fortune by forecasting the weather. When his forecast suggested a good olive harvest, he bought olive presses while they were cheap and resold them at a much higher price when the good harvest duly arrived—and olive presses became in short supply. It's said that he did this in response to people who said that there was no point in studying what we would now call science.

More reliable stories say that he successfully predicted a solar eclipse and was an engineer and businessman. Today, a vast number of professional mathematicians, including many physicists, economists, and engineers, spend their time developing theorems and trying to prove or disprove them.

Types of proof

There are five main types of proof in mathematics, and almost every mathematical theorem can be proven using one or more of them.

1. Direct Proof

This is the kind of proof that is used most often. It proceeds by a series of steps. To prove that "A implies B," you might be able to start like this:

"I know that A implies C."
"I also know that C implies B."
"Therefore, A implies B."

Here's an example.
Theorem: If n is an even number, then n^2 is also an even number.

Proof: The definition of an even number is that it can be divided by 2 to give a whole number. So, 10 is even because dividing it by 2 gives the whole number 5. Odd numbers such as 5 always divide into a fraction (5 ÷ 2 = 2.5) but multiplying it by 2 will get back to a whole number.

This definition means that any even number can be written as 2w, where w is a whole number.

Our theorem says n is an even number, so
n = 2w

Pressing olives to release the valuable oil was hard work in ancient times, but Thales found a way to make it pay.

How it works

Proving Thales' theorem
The theorem named after Thales states that, if one draws a triangle so that it fits into a circle, and one of its sides is the circle's diameter, then the opposite corner will always be a right-angle.

Thales' proof is based on two facts:
1. The sum of the angles in a triangle equals two right-angles (180°).
2. In an isosceles triangle, two of the angles are always the same.
The theorem is proved by drawing a triangle ABC in a circle, then drawing a line from the center of the (O) to the corner B of the triangle, which divides the triangle into two smaller ones. These are both isosceles triangles.

Because the new triangle AOB is isosceles, we know that two of its angles, labeled α (which is the symbol for alpha, the Greek letter "a") are equal. And, two of the angles in the other isosceles triangle BOC are also equal and are labeled β (for beta).

This is where the algebra comes in. We know that the internal angles of ABC add up to 180°, so from the diagram, we can see that:

α + (α + β) + β = 180°

So

2α + 2β = 180°

Which means

2(α + β) = 180°

And dividing both sides by 2 gives us the proof we are looking for:

(α + β) = 90°

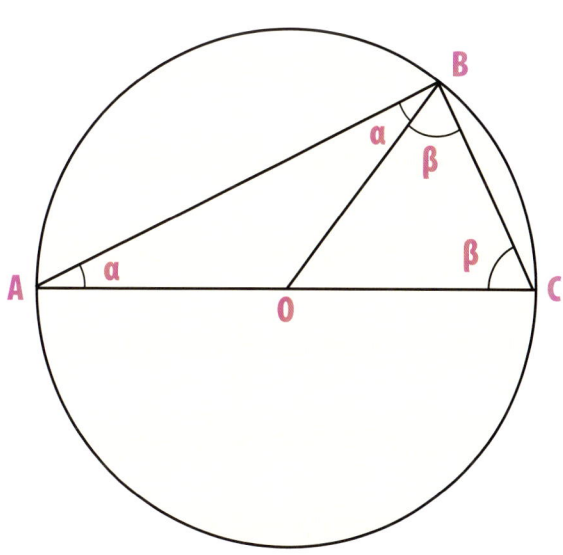

Let's square this equation, to give
$n^2 = (2w)^2 = 4w^2$

This can be written as
$n^2 = 4w^2 = 2 \times 2w^2$
(We'll see why this is useful in a moment)

Now, let's define a new number
$m = 2w^2$

So now we can say that
$n^2 = 4w^2 = 2 \times 2w^2 = 2m$

Now we can go back to our definition again and see that 2m must be an even number, because, when divided by 2, it gives m, which is a whole number. Finally, we can say that n^2 must be even, too, because it is equal to 2m, which we have just proved is even.

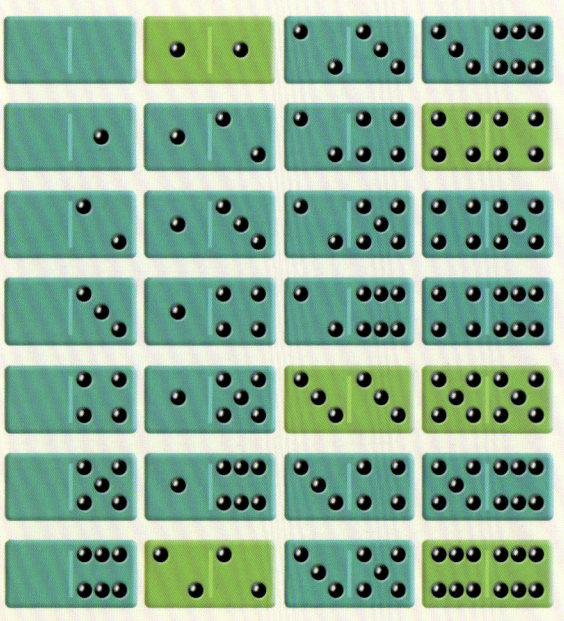

One thing to notice here is that you have to be creative to solve a proof. Deciding to square the equation in the first step turned out to be a useful idea, but in other cases it might not have helped, and there are other things that might have worked just as well here. Choosing what to try is a matter of experience (or good luck), and it is what makes math fun!

2. Proof by Induction

To prove a theorem about a series of numbers, proof by induction is often used. It is based on the idea that, if you can select any number from a series and prove that the theorem is right about that one, and that the theorem is also always right about the following number in the series, then it is true for all numbers in the series.

For instance, to prove
$1 + 2 + 3 \ldots + n = n(n+1)/2$

That is, to show that the equation is correct, whatever value of n is chosen, first we prove it is true for a selected value of the number n.

Let's choose n = 2
$1 + 2 = 2(2 + 1)/2$

Which gives
$3 = 6/2$

The double dominoes always add up to an even number, even if they are showing odd numbers.

The Sun rose yesterday, so it will rise tomorrow. That's induction. However, in this picture it has been eclipsed as well. In 585 BCE, Thales, the first mathematician, was the first person to predict a solar eclipse in advance.

So the theorem is correct for 2. We could easily prove that it works for the next number, 3, but we want to be able to prove that it works for a selected number, and the next number whatever that selected number is. That is, we must show that if it works for any number (let's call it k), it must also work for the next (k+1).

We have just shown that
$1 + 2 + … + k = k(k+1)/2$
for at least one value of k (that is, 2). Now, let's add k + 1 to both sides of that last equation

$1 + 2 + … + k + k + 1 = k(k+1)/2 + k + 1$

and then rearrange the right-hand side: First, we put brackets round it, multiply everything inside by 2, and then divide the whole thing by 2:

$1 + 2 + … + k + k + 1 = (k(k+1) + 2k + 2)/2$
$1 + 2 + … + k + k + 1 = (k^2 + k + 2k + 2)/2$
$1 + 2 + … + k + k + 1 = (k^2 + 3k + 2)/2$
$1 + 2 + … + k + k + 1 = (k + 1)(k + 2)/2$

Which shows that our original equation,
$1 + 2 + 3 … + n = n(n+1)/2$
works both for k and for k + 1.

Christopher Columbus sailed for India, and so that is where he must have arrived, he thought. "Look, Indians! No further proof necessary!"

"If we are in France, then we are in Europe."
"If we are not in Europe, then we cannot be in France."

"If you have the pencil, then you have a writing tool."
"If you don't have a writing tool, then you don't have the pencil."

These pairs of statements all have the pattern:

"If A then B."
"If not B, then not A."

As k can be any number, this shows that our equation works for all numbers, which is what we wanted to prove.

Something to notice here, which is true of all proofs, is that they rest on other parts of mathematical knowledge. For instance, in the above proof we go from $k(k + 1)$ to $k^2 + k$, which requires us to know how multiplication works with brackets, and also that k times k is k^2.

3. Contrapositive Proof

The idea behind contrapositive proofs can be hard to get your head around. It is based on a logical pattern that is shared by pairs of statements, like these:

"I am human, therefore I am a mammal."
"I am not an mammal, therefore I am not human."

And by reading them we can see that these two statements come to the same thing. So, if we can prove one of them, the other must also be true. The second statement is called the contrapositive of the first.

Like all these proofs, it's easiest to see how they work by using an example:
"If n^2 is even, then n is even."

Let's look at the contrapositive statement, which is:
"If n is not even, then n^2 is not even."

If we can prove this is true, then the original statement is true, too.

"Not even" means "odd," so we can rewrite our contrapositive as:

"If n is odd, then n^2 is odd."

Any odd number can be written as
$n = 2k + 1$
where k is an integer; that is, a whole number in the series ... -2, -1, 0, 1, 2, ..., so for instance, the odd number 9 can be written as $9 = 2 \times 4 + 1$.

What can we say about n^2? We can find out by squaring both sides of that last equation
$n^2 = (2k + 1)^2$
$n^2 = 4k^2 + 4k + 1$

We can rearrange this, so that it looks like the equation for an odd number, like this
$n^2 = 2(2k^2 + 2k) + 1$

The part in brackets here, $2k^2 + 2k$, is made up of integers, which means that it must be an integer itself. And that means our last equation says exactly the same as the equation for n above:
$n = 2k + 1$

That must mean that n^2 is also an odd number.

So, we have proved that, if n is odd, then n^2 is also odd. And, since this is the contrapositive of our original statement, then we have proved that statement, too.

Contrapositives are two opposite statements that use the same logic.

4. Proof by Contradiction

This kind of proof works by showing that a theorem must be true because assuming that it isn't leads to a contradiction. Here's a famous example of this kind of proof:

If we leave out the number 1, then whole numbers are either "composite," which means they can be formed by multiplying together other whole numbers, or "prime," which means they can't be. So, 5 is a prime number because it can't be produced by multiplying any other numbers together. But 6 is the product of 2 and 3, so it is a composite number. And, in the end, any composite number can be made by multiplying primes together. For instance, $24 = 4 \times 6 = 2 \times 2 \times 2 \times 3$. That is to say, every composite number has a unique set of prime factors.

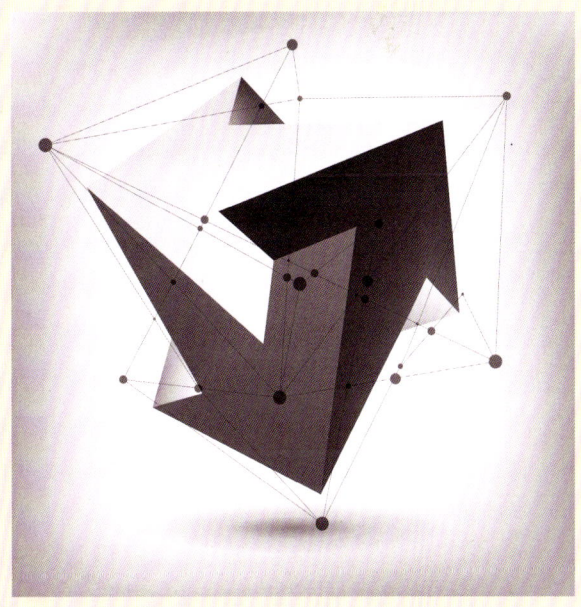

Euclid, an influential ancient Greek mathematician, proved 2,300 years ago that there is an infinity of primes. He is seen here sketching out a math problem for students to see. Euclid did not prove the primes problem quite as set out below using algebra. Instead he did it—by contradiction—using geometry, that compared the lengths of lines. Euclid included the proof in his book, *The Elements*, which has never been out of use since he wrote it. Copies are still produced today.

The theorem we want to prove is:
"There is an infinite number of prime numbers."

And this is how we do it. We assume that the opposite to our theorem—there is a finite number of primes—is true and show that this leads to a contradiction:

1. If there is not an infinite number of prime numbers (p), then we could multiply them all together. Let's do this, add 1 to the list, and call the result N. That is:

$$p_1 \times p_2 \times p_3 \times ... \times p_n + 1 = N$$

What can we say about N?

2. N can't be a prime, because we've used up all the prime numbers there are, and N is bigger than any of them.

3. So N must be composite. In other words, it must have prime factors.

4. We've listed all the prime numbers already, so N's prime factors must be in that list.

5. But in order to get to N, we can't just multiply prime numbers together, we also need to add 1 to them.

6. So N can't be made just by multiplying prime numbers together.

7. So N can't be composite.

8. We've already shown that N can't be prime, and now we've shown that it can't be composite either. That is a contradiction, so our original assumption, that there is not an infinite number of primes, must be wrong.

9. So, there is an infinite number of primes.

5. Proof by Example

This method of proof, which is also known as proof by construction, is very simple to apply, though there are not many theorems that it can be used for. Here's an example:

"All primes are odd."

This can be disproved simply by noting that 2 is a prime which is not odd. So, 2 is the example that disproves the theorem.

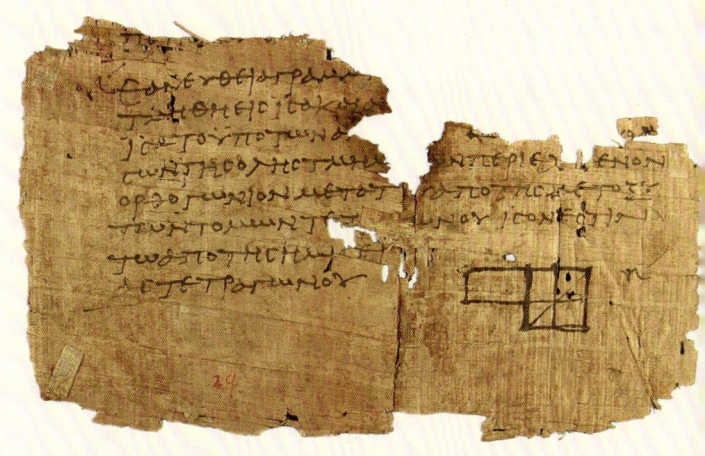

Above: The oldest surviving version of Euclid's *The Elements* is a scrap of papyrus from the 1st century CE.

Left: The highlighted numbers are the primes below 100. They all look to be odd, because all even numbers are multiples of 2. However, right at the start the very first prime shows that not all primes are odd.

SEE ALSO:
▶ The Fundamental Theorem of Algebra, page 130

The Pythagoreans

PYTHAGORAS'S THEOREM IS SURELY THE MOST FAMOUS MATHEMATICAL FORMULA OF ALL, and one of the oldest, too. In fact, it was known long before Pythagoras himself was born. He was, however, probably the first to prove it.

Pythagoras was undoubtedly a brilliant man, but he also had the good fortune to be born in one of the greatest mathematical cultures of his time, and to spend time in the other two great centers of math. He was born around 570 BCE on the Greek island of Samos in the Aegean Sea. In his twenties, he traveled to Egypt, where he spent the next 22 years meeting every mathematician and astronomer he could track down. Then, in 525 BCE, the army of the Persian king Cambyses II invaded Egypt and Pythagoras was taken prisoner. We don't know how he turned this situation to his advantage, but he managed to continue his mathematical studies in Babylon, somehow gaining access to the greatest thinkers there as well. Twelve years later, he returned to Samos. Then, at the age of about 56 (an old man by the standards of his time), he was ready to change the world.

Musical math
And so he did, starting with what might at first have seemed a trivial discovery. Pythagoras found that, if two strings of a musical instrument are struck, then the sounds they make will go well together if one string is twice the length of the other (assuming that they are the same

The modern harbor at Samos has a statue of the island's most famous son, equipped, of course, with a right-angled triangle.

Spending his early adult life in ancient Egypt, Pythagoras came into contact with many different schools of thought.

thickness and tension, and made of the same substance). Today, we call this harmonious pair of sounds an octave. Pythagoras also found that if one string is two-thirds the length of the other, or three-quarters, or four-fifths, the resulting double sound will be harmonious as well. On the other hand, if the ratio of the lengths is not a simple fraction, the pair of sounds will be out of tune.

All is number

So impressed was Pythagoras by this finding that he decided that numbers are the most important things there are. In fact, he said that the whole universe is made of them. By "numbers" he meant positive whole numbers, those in the series 1, 2, 3 …, and also fractions involving those numbers, like those he had found to be so important in producing music. Such fractions are also called ratios, and so we call the numbers involved rational numbers. The importance of them to Pythagoras and the scientists who followed him is why today the word "rational" means "logical" or "sensible." Pythagoras was determined to develop a whole system of

Pythagoras, by now an old man, instructs his followers on a proof of his famous theorem about right-angled triangles.

mathematics based on rational numbers and on Thales' concepts of theorems and of proofs.

The move to Italy

In about 530 BCE, Pythagoras moved to Croton in what is now Italy (but was then part of the Greek empire), where he attracted a number of students to whom he taught mathematics. Unusually for the time, both men and women were allowed to join. Together, the Pythagoreans began to work on Pythagoras's grand scheme to unlock the mathematical secrets of the universe. But they were not just a mathematical research team.

They worshiped numbers, and invested them with all sorts of sacred powers and magical properties (see box, page 31), and they developed a whole religion in which numbers played an essential role. They kept the details a secret, so we know only some scattered facts about their beliefs (see box, page 33).

Not rational at all

Tragically for Pythagoras, his great mathematical project soon ran into trouble through the very theorem that we remember him by. Pythagoras's theorem refers to right-angled triangles, and the simplest such triangle is one whose short sides are the same length. If these sides are 1 unit long (cm, cubit, inch—the units don't matter), then the length of the longer side, the hypotenuse, is

$$a=\sqrt{(1^2+1^2)} = \sqrt{(1+1)} = \sqrt{2}$$

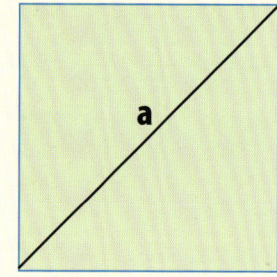

A square is two simple, right-angled triangles. The length of the diagonal of any square cannot be expressed as a rational number, despite Pythagoras's suggestion that nature, order, and beauty could be explained using only whole numbers.

Having got this far, the next step for a Pythagorean would have been to work out what number √2 is—that is, what number, when squared, gives the answer 2. But try, as they

How it works

Proving Pythagoras

There are over a hundred proofs of Pythagoras's theorem. Here is one of them.

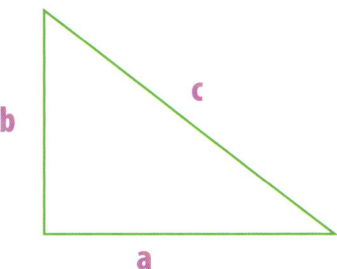

We start by working out the area of a right-angled triangle. We could make a triangle like the one above by drawing a rectangle with sides a and b and cutting it in half along the diagonal. The area of that rectangle would have been ab, so the area of half of it is ½ab.

Now, take 4 such triangles. Their total area is 4 x ½ab, which is 2ab. Arrange them in this pattern:

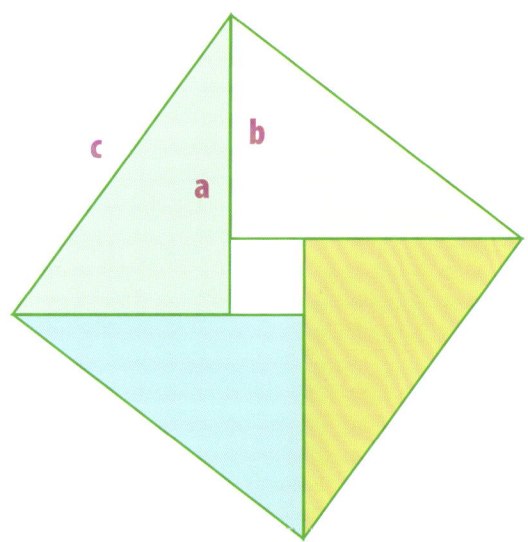

The vertical line toward the left of the figure is the base of one of our triangles, so its length is a. The upper part of that line lies next to the shortest side of another triangle, so the length of that upper part is b. So, the length of the lower part of the line must be a – b.

This is one side of the square in the middle of the figure. So, the area of that small square must be $(a-b)^2$.

So, that means the area of the whole figure (4 triangles and the square) is

area = 2ab + (a – b)²

But, each side of the figure is a hypotenuse of one of our triangles, which has a length c.

So we can also express the area as

area = c²

Putting these two formulas for the area together:

2ab + (a – b)² = c²

Multiply this out:

2ab + a² – 2ab + b² = c²

And simplifying gives

a² + b² = c²

This is Pythagoras's theorem.

might, the Pythagoreans failed, for the very good reason that there is no such number. That is, no number that can be expressed as a ratio, which, according to the Pythagoreans, are the only kind that exist. According to some accounts, it was a Pythagorean called Hippasus who discovered that √2 is not a rational number, and in punishment the cult deliberately wrecked the ship he was on in order to drown him.

Beyond use

Although we accept the existence of such numbers as √2 today, and simply refer to them as irrational numbers to distinguish them from the rationals that Pythagoras preferred (see box, page 32), the discovery really was astonishing.

Algebra had been developed for very practical uses, including surveying, which is based entirely on measurements and how to manipulate them: How to calculate the area of a plot of land from its shape and the measured length of its perimeter, for example. The discovery of irrational numbers meant that it wasn't always possible to make such measurements properly. If you have a right-angled triangle with the shorter sides being 3 units and 4 units long (again, what these units are doesn't matter), you can measure the hypotenuse to be 5 units long. But, what if the short sides are the same length, say 100 inches long? How do you measure the hypotenuse then? If you measured the sides by using a ruler marked in inches, you could try measuring the hypotenuse with that same ruler. But it would only tell you that the hypotenuse is

Ancient Egyptian surveyors, known as rope stretchers, used a cord with 12 equally spaced knots. This could be arranged into a triangle with sides of 3, 4, and 5 knots. The triangle would always have a right angle, which is ideal for plotting the corner of a field.

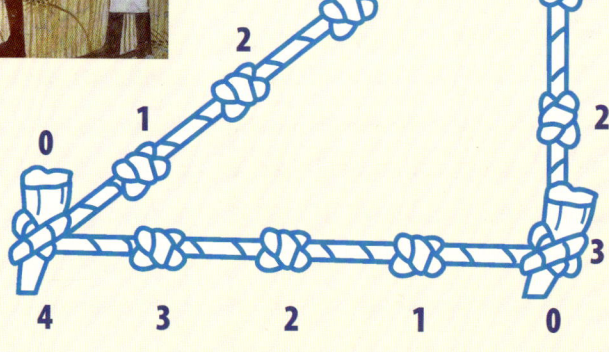

MAGICAL NUMBERS

Many ancient cultures, and some modern ones too, believe that numbers have a sacred significance. In fact, even today, many people believe that 13 is unlucky and may have their own personal lucky number. People who take such things seriously are sometimes called numerologists. They believe, for example, that everyone's name can be translated into a number, and that this number holds secrets about the person. But it isn't likely that any mathematicians today are also numerologists. To the Pythagoreans, however, mathematics and numerology were just two ways of studying numbers, each taken as seriously as the other. For them, the most sacred and magical number of all was 10, particularly when it referred to a triangular pattern of 10 dots which they called the "tetractys of the decad."

You could keep using more and more accurate rulers, but however good they were, you would never be able to measure the length exactly. Even if you had a ruler marked in ten thousandths of an inch, and used a microscope to read it off, you would still only be able to say that the length was between 1,414,213,562 and 1,414,213,563 ten thousandths of an inch.

No way out

You might think you can get round this by using the hypotenuse itself as a ruler. If you tried that, marking off a ruler to be the same length as the hypotenuse, dividing its length by, say 1,000, and then measuring the sides in these new units, you would find the same problem. Your ruler, while giving you an exact length of 1,000 for the hypotenuse, could only tell you that the shorter sides were both between 707 and 708 units long.

No ruler can help. Some lengths simply cannot be measured accurately.

between 141 and 142 inches long. But what is the exact length? You could get a more accurate ruler, one marked in tenths of an inch, and try that. You would then discover that the hypotenuse is between 1,412 and 1,413 tenths of an inch long. Still not exact.

TYPES OF NUMBER

Whole numbers
Numbers in the sequence
0, 1, 2, 3 …

Counting numbers
Numbers in the sequence
1, 2, 3 …
Some authorities also treat 0 as a counting, or natural, number.

Integers
Numbers in the sequence
… −3, −2, −1, 0, 1, 2, 3 …

Rational numbers
Numbers which can be expressed as ratios (or fractions), such as:

… −$^{12}/_5$, −2, 0, $^1/_8$, $^2/_3$ …

Irrational numbers
Numbers which can't be expressed as ratios, such as:

−$\sqrt{3}$, $\sqrt{2}$, $\sqrt[3]{2}$, π

Transcendental numbers
Some numbers can be expressed as the solution (also known as the "root") of a polynomial equation. For instance, the solution of the quadratic equation $x^2 = 2$ is $\sqrt{2}$. But there are numbers which are not the solution of any polynomial, such as π. These are called transcendental numbers, because they transcend (go beyond) the range of all other numbers.

Since the $\sqrt{2}$ problem undermined their entire belief-system, the Pythagoreans tried to keep the discovery secret and forbade any of their members from discussing it. But the terrible truth soon became well-known and as a result, the number-based approach of the Pythagoreans to all natural processes fell from favor.

Violent end

Nevertheless, the leading members of Crotonian society wanted to join the group, making it a very powerful organization which controlled much of the city. One of the leading Pythagoreans was a wrestling champion called Milo, who had rescued Pythagoras from a collapsing building, and whose house was one of the main meeting-places of the cult. When war broke out between Croton and the nearby city of Sybaris in 510 BCE, it was Milo who led the Crotonians into battle with them. The Crotonians defeated the Sybarites, making the positions of Milo, Pythagoras, and the Pythagoreans even stronger. Cylon, a wealthy and powerful man, and also a brutal bully, asked to join the cult, but because of his reputation the Pythagoreans rejected him. But their decision was to have fatal consequences. Cylon and his friends argued that Croton should not be controlled by the cult, but should instead be a democracy. Antagonism between the groups led to violence, and in 508 BCE Pythagoras fled to the city of Metapontum, where he spent the rest of his life. Some of the Pythagoreans were killed, but many escaped to other Greek cities and set up their own groups. Milo, meanwhile, ended up being eaten by a pack of wolves!

THE RULES OF PYTHAGORAS

Although we don't know much about the beliefs of the Pythagoreans, a set of the rules by which they lived their lives has come down to us—and they sound very strange indeed. Presumably, they are designed to fit some theory about the nature of the universe, but as we have no idea what that was, we can't make sense today of the rules' purpose.

- ▲ To abstain from beans
- ▲ Not to pick up what has fallen
- ▲ Not to touch a white cockerel
- ▲ Not to break bread
- ▲ Not to step over a crossbar
- ▲ Not to stir the fire with iron
- ▲ Not to eat from a whole loaf
- ▲ Not to pluck a garland
- ▲ Not to sit on a quart measure
- ▲ Not to eat the heart
- ▲ Not to walk on highways
- ▲ Not to let swallows share one's roof
- ▲ When the pot is taken off the fire, not to leave the mark of it in the ashes, but to stir them together
- ▲ Not to look in a mirror beside a light
- ▲ When you rise from the bedclothes, roll them together and smooth out the impression of the body

SEE ALSO:
▶ The Algebra of Shapes, page 34

The Algebra of Shapes

THE DISCOVERY THAT NOT ALL NUMBERS ARE RATIONAL caused a loss of faith in numbers as a trustworthy way of getting at the truth. So, in response, Greek mathematicians turned back to the older geometrical methods of the Babylonians.

The greatest by far of this new generation of Greek mathematicians was Euclid. Translations of *The Elements*, his multi-volume geometry work, were being used to teach geometry in some schools right up to the early 20th century, more than 2,000 years after his death.

Using a system

The Elements marked the birth of a new kind of algebra, now called algebraic geometry, and it was also a milestone as the first example of a system in mathematics. A mathematical system is one in which a whole range of theorems is proved from a small number of carefully formulated and unchallengeable assumptions called axioms. For example, one of Euclid's axioms could be written as "if a = b and b = c, then a = c." Ever since Euclid, great mathematicians have followed in his footsteps in trying to develop systems based on stated assumptions and locked together by rigorous proofs.

Mystery figure

Considering the importance of Euclid and the fact that so much of his work survives, we know surprisingly little about him. Unlike other ancient Greek mathematicians, no one seems to have written a biography of Euclid, and the few details

No one knows what Euclid looked like—or even if he was an actual person—but he is invariably shown with a beard.

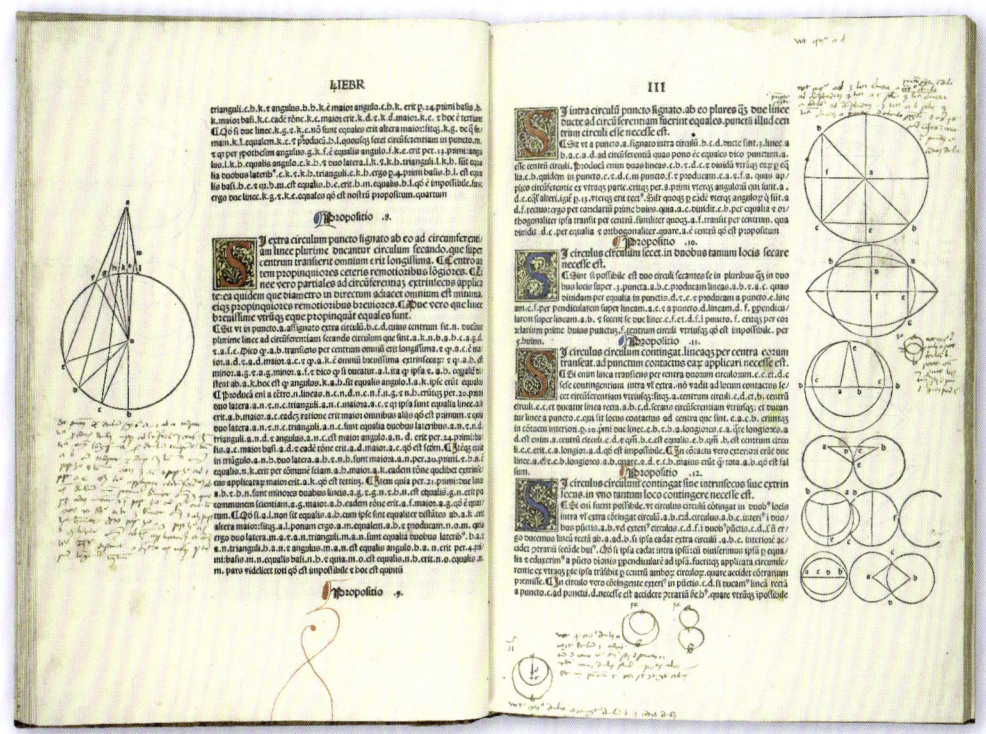

Euclid's *The Elements* is the most successful non-religious book in history. It has never been out of use since its first publication 2,300 years ago.

we have about him were recorded five centuries after his death. In fact, even his name was rarely used by later Greeks. Instead, he was often just referred to as "the writer of *The Elements*." We do know, however, that Euclid lived and studied in Alexandria, Egypt and had access to the greatest repository of knowledge of the ancient world, the Great Library of Alexandria. The Library may have held almost half a million books, most in the form of scrolls of papyrus (a type of paper made from reeds, not wood pulp). There, Euclid would have been able not only to refer to earlier mathematical texts, but also to work with other scholars and researchers.

Making shapes

Even today, there is something very appealing about using diagrams to solve mathematical problems (see box, overleaf), and Euclid's system was highly successful. It was widely believed, in fact, that any mathematical problem could be solved using just two drawing tools: A compass and a straight edge—not a ruler marked with units; no Greek mathematician would have wanted to rely on a tool they could not be certain of at a glance. In fact, there were only three major problems which resisted all the attempts of Greek mathematicians to solve them using straight edge and compass (see box, page 39).

The Library of Alexandria collected knowledge by making copies of every book that arrived in the bustling port city.

Modeling nature

Although the Greeks were interested in mathematical studies for their own sake, they were also well aware of great practical benefits of algebraic geometry in subjects like architecture and astronomy. Euclid himself wrote books about astronomy and what we would now call geometrical optics—that is, using geometry to solve practical problems involving light. For many centuries after Euclid's death, algebraic geometry was the most powerful mathematical tool available to scientists. Galileo, one of the greatest scientists, used it to explore and test his theories and, like the Greeks themselves, believed that the Universe was actually constructed according to geometric principles. In 1623, he wrote that what we would call science "… is written in that great book which is always visible to our eyes—I mean the universe—but we cannot understand it if we do not first learn the language and grasp the symbols in which it is written. The book is written in the language

GEOMETRIC PROOFS

Often in algebra, we wish to prove an identity, where two expressions are always equal. For example, is this an identity?

$$(a + b)^2 = a^2 + b^2 + 2ab$$

The Greeks were able for the first time to prove such identities. They used geometry to do so.

To prove the one above, all that is needed is a rectangle; any dimensions will do. If the lengths of its sides are a and b then its area will be ab.

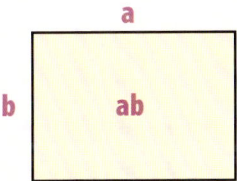

Now, draw a square on the short side of the rectangle, and another on the long side. The areas of these two squares will be a^2 and b^2.

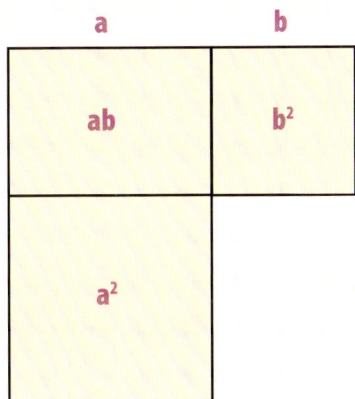

Finally, add in another rectangle to make a large square. This rectangle's sides are of length a and b, so its area is ab.

So, the total area of the large square is, as we can see from the final figure,

$$\text{area} = a^2 + b^2 + ab + ab$$

Which is

$$\text{area} = a^2 + b^2 + 2ab$$

But we can also see that the large square has sides of length a + b. So, its area can also be written as

$$\text{area} = (a + b)^2$$

Putting together these two ways of writing the area give us

$$(a + b)^2 = a^2 + b^2 + 2ab$$

Which is the identity we wanted to prove.

Galileo Galilei was the first scientist to present laws of physics in terms of math.

of mathematics, and the symbols are triangles, circles, and other geometrical figures, without whose help it is impossible to comprehend a single word of it; without which one wanders in vain through a dark labyrinth."

New techniques

Algebraic geometry only lost its popularity once calculus was developed in the 17th century, and even Isaac Newton, who studied gravitation and motion using calculus, still used geometry to prove his results when he published them.

SEE ALSO:
▶ Proof, page 16
▶ Calculus, page 110

Galileo used this geometrical diagram to study his theorem that the time taken by an accelerating object to move a particular distance is the same as that taken by an object moving at a steady speed, if that speed is half the top speed of the accelerating object. Vertical lines are times, horizontal ones are speeds, and areas represent distances (because distance = speed × time). He used geometry to show that the area of the rectangle ABFG is the same as the area of the triangle ABE, which proves his theorem.

THE THREE CLASSIC PROBLEMS

The ancient Greeks were phenomenally successful at solving problems by using geometrical techniques that required only a compass and a straight edge, but there were three problems which they could not crack, despite many attempts to do so. It has since been confirmed that they are impossible to prove by compass and straight edge.

1. The Delian problem

According to legend, when a plague struck the Greek island of Delos, the citizens asked the oracle of the god Apollo for help. The oracle told them to build a new altar twice the size of the existing one, which was a perfect cube. Accordingly, a new altar was built, with sides and height twice that of the original, but the plague continued. So, they realized that what was needed was not an altar with double-length sides, but one with a double volume. All they had to do was work out how long its sides would be, and they could build it. Accordingly, they reached for their compasses and straight edges ... Using modern notation, this problem becomes: Given a cube with sides of length a (and with a volume that is therefore a^3), calculate the length, x, of the sides of a cube with volume $2a^3$.

The formula for this problem is $x = \sqrt[3]{2a^3}$ which, as you see, involves the roots of 2, which just will not fit into everyday sums.

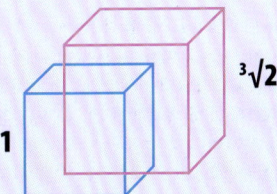

2. Squaring the circle

The problem here is to find a square with exactly the same area as a circle. Again, this is very easy to express using modern algebra:

$x^2 = \pi r^2$, so $x = \sqrt{\pi r^2}$

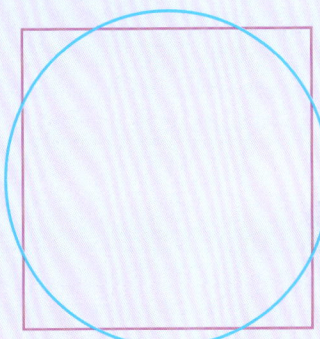

However, now, as then, it is not possible to find an exact value for x, because π is transcendental, which means not only that it can't be written down exactly, but that it can't even be represented as the solution of a polynomial equation. We can approximate it as closely as we wish, though.

3. Trisecting an angle

For the ancient Greeks, this must have seemed temptingly possible, because they did know how to bisect (halve) an angle using a straight edge and compass, but a way of dividing it into three proved elusive.

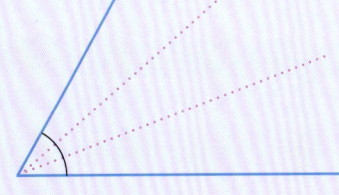

The Prehistory of Calculus

CALCULUS IS PROBABLY THE SINGLE MOST POWERFUL MATHEMATICAL TOOL SCIENCE HAS, because it is the only way to properly understand change. Change is what science is all about, from swooping helicopters to exploding atoms, and from shifting continents to the expanding universe.

Although the ancient Greeks had little interest in change, and their scientific studies were mainly concerned with shapes and patterns, they did develop some ideas which would later be taken up and turned into calculus, once science had progressed to the point where it was required to handle changing phenomena. This is because calculus is also a way of moving from one kind of space to another, which is something that the Greek thinkers were fascinated by.

Adding dimensions

Take a simple shape like a line, joined end to end to make a circle. Today, we could write the length of that line as $2\pi r$, where r is the radius of the circle, and is measured in units of length,

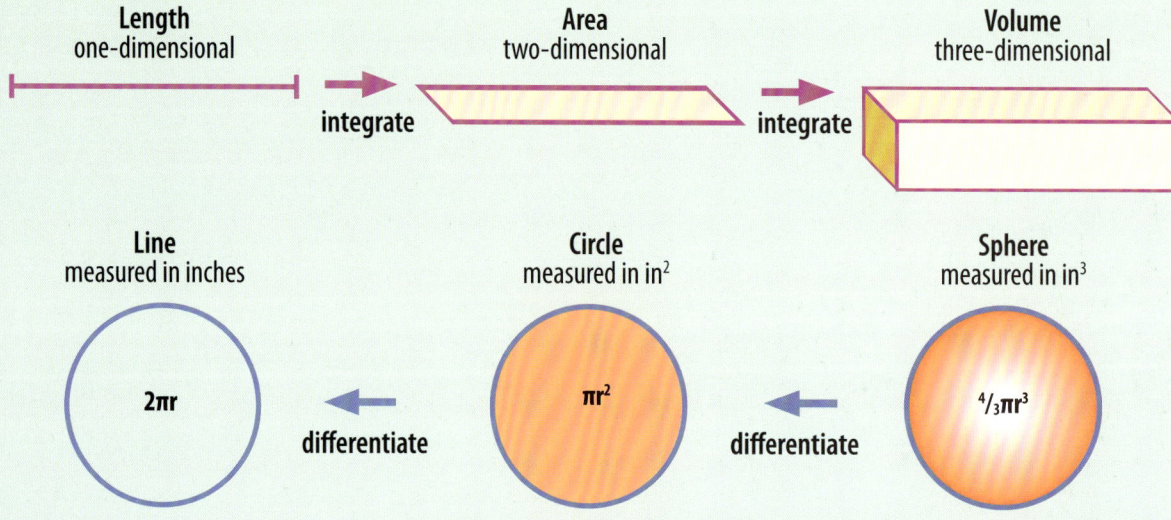

such as inches (in). What about the area of the circle? Its formula is πr^2, and it is measured in units of area: in^2. What is the three-dimensional equivalent of a circle? A sphere, with a volume measured in cubic in (in^3), and given by $4/3 r^3$. What's this got to do with calculus? Well, as we'll see on page 110—and illustrated below left—if we differentiate the expression for a three-dimensional object, we get the expression for a two-dimensional one, and if we then differentiate that, we get to a one-dimensional formula. And integration takes us in the opposite direction.

Cutting cones

Rather than dealing with general formulas like πr^2, calculus can also be used to analyze a particular shape, for example to find the area it encloses. These were challenging concepts for the ancient Greeks and, while a few of them made discoveries about how one might estimate the length of a curve by replacing it with a series of joined straight lines, it was Archimedes who made real breakthroughs in this area. For instance, he was interested in conic sections, which are the lines one gets if one cuts a cone at different angles.

Curved lines and straight shapes

One of these sections is the parabola, and Archimedes wanted to know what the area under it is. The parabola looks a little like a triangle, so Archimedes would have started by fitting a triangle into it. The area of a triangle is ½bh, where b is base and h is height, so, if b is 2 feet and h is 3 feet, its area is: **½ x 2 x 3 = 3 ft²**.

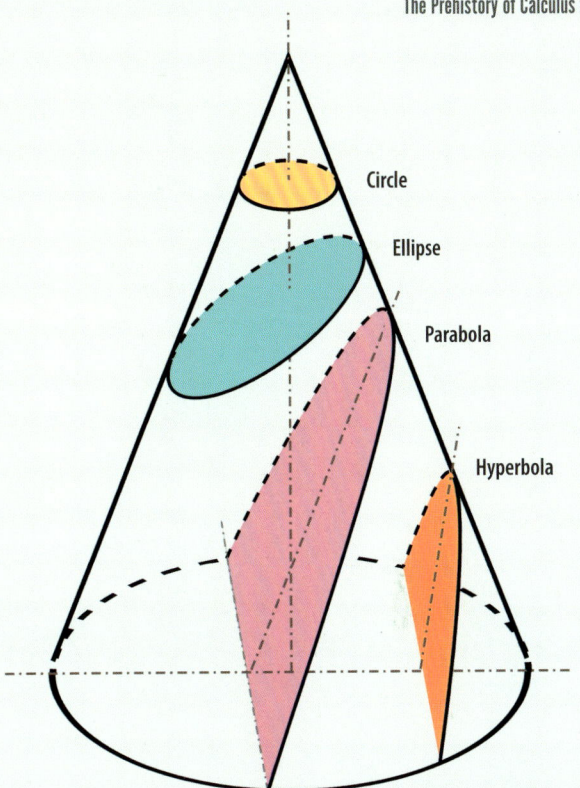

Cutting a three-dimensional cone at different angles results in four types of two-dimensional curve.

But the parabola clearly encloses an area larger than this. So, Archimedes' next move was to fit in more triangles, work out their areas, and add them to the total. And he could continue doing this, until he got as close to the area under the parabola as he wanted. This is similar to the way in which integral calculus works, by dividing areas under curves into many tiny areas which are then added together. Archimedes used a similar method to calculate a good estimate of π (see box, overleaf).

Going the other way

Archimedes took the first step in the other direction, too, toward differential calculus, when

Archimedes is said to have invented an armory of incredible weapons that saved his city from invasion by a Roman fleet.

he developed a technique to find the slope of a curve at any selected point. This is one of the most important applications of differential calculus, although the modern method is very different from that used by Archimedes.

Pure thought

Archimedes was certainly the greatest mathematician of the ancient world, and one of the greatest of all time; like other great thinkers he developed completely new ideas in more than one area (including a version of what we would call cubic equations; see box, overleaf), and some of them were really too advanced to be understood properly by his contemporaries. Archimedes was the son of an astronomer, and was either a relative or a friend of Hieron, the king of Syracuse (a Greek colony in Sicily). Archimedes was born there in 287 BCE. Like all other famous ancient Greek thinkers, Archimedes was interested in what we would call pure, rather than applied, science. The Greeks believed that thinking and discussion were the best ways to advance knowledge, and had little interest in experimentation or measurement. In this way they were very different from the Babylonians, who valued mathematics only as a useful tool for farmers, accountants, and builders. Actually, the idea that pure science (and especially pure mathematics) is somehow better than the applied versions survived well into the 20th century.

Bath-time breakthrough

Archimedes' close friendship with Hieron led him to carry out a number of practical scientific studies. The most famous is his method of working out the purity of Hieron's supposedly gold crown by measuring its density. When he hit on the answer, which was to weigh the water the crown displaced, he is said to have run down

ARCHIMEDES' ESTIMATE OF π

To calculate π, we can start by drawing a circle, and drawing one square inside it and another outside it.

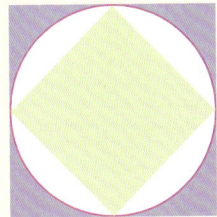

Let's say the circle is 1 unit across (the unit doesn't matter—it could be an inch, a centimeter, or a mile across). As the circle just fits across the outer square, the sides of that square must be 1 unit long, too. The total length of the sides of the square must be 4. Let's call this the perimeter of the square. The circle fits inside the square, so the circle's circumference (the length of which is equal to π) must be less than that of the square. Which is to say that π is less than 4.

If you look at one side of the inner square, you can see that it forms the hypotenuse of a right-angled triangle, with the two shorter sides each being 0.5 units long. By Pythagoras's theorem, we know that this means that its length is $\sqrt{(0.5^2 + 0.5^2)}$ which is $\sqrt{0.5}$.

So, the perimeter of the smaller square is $4\sqrt{0.5}$, which is about 2.828. From this we know that π is greater than 2.828. Writing what we found more briefly:

2.828 < π < 4

This isn't a very accurate approximation! But then, the circle is a very different shape to a square. To increase accuracy, Archimedes repeated the approach with figures of more and more sides, achieving a closer and closer fit to the circle, and so a more and more accurate value of π:

One of the advantages of this method is that there is no need to measure anything; the only point of the diagrams is to illustrate what is going on. The disadvantage is that it becomes more and more complicated to work out the perimeters of the shapes. For the six-sided figure, for instance, we would calculate the lengths of the sides by trigonometry

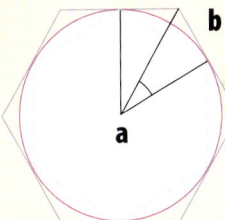

The task was much harder for Archimedes, because this kind of trigonometry had not been invented in his time, so he had to use much clumsier geometrical methods to work out his answers. So it is a great tribute to his determination, as well as his skill, that he went right up to 96-sided figures!

This gave him a value of π of 3.1418, which is just one-hundredth of 2 percent from the correct answer: 3.14159265 …

CUBIC EQUATIONS

Archimedes lived long before modern algebraic notation, and he solved his version of a cubic equation by using geometry. Cubic equations have a degree of 3, and three answers. Today, we would write the general form of a cubic equation like this:

$$y = ax^3 + bx^2 + cx + d$$

If we set $a = 2$, $b = -1$, $c = -1$, $d = 0$, we get a particular cubic equation:

$$y = 2x^3 - x^2 - x$$

And if we put a range of values in for x, we obtain the following plot:

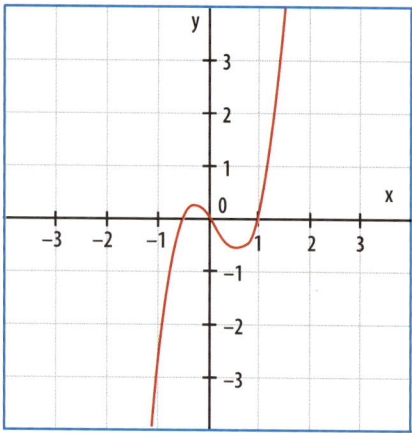

The solutions (or roots, or zeros) of the equation are the points where $y = 0$, which are the places where the line crosses the x-axis. These are -0.5, 0, and 1.

the streets naked shouting "Eureka!" (meaning "I have found it!"). Although this sounds an unlikely story, it is said that he would often write calculations on his own oily body after a bath, using ash from the fire, so maybe he really did have a bath-time breakthrough.

Archimedes was murdered by a Roman soldier who flew into a rage when the great thinker did not follow orders.

Math as a weapon

Archimedes is also said to have demonstrated to Hieron that math can be useful as well as interesting, by building a compound pulley in the harbor. When it was attached by rope to a fully-laden ship, Archimedes used the pulley to drag the ship across the water single-handedly. This particular invention proved that math could be far more useful than anyone could have imagined. Syracuse was often at war with the growing Roman Empire and, in 212 BCE, was besieged by a fleet of well-armed Roman galleys. Archimedes, by now 70 and a very old man indeed by the standards of his time, helped defend his city by using cranes with iron claws attached to his compound pulleys to drag some of the galleys to destruction. His pulleys may have used levers, and Archimedes is famous for describing the power of leverage, saying, "Give me a lever long enough and somewhere to stand and I will lift the world." He also invented a deadly new kind of ship-destroying catapult, and may even have used curved mirrors to focus the Sun on the sails of galleys, setting them on fire.

Murdered for math

Thanks to Archimedes, the Roman attack was repelled. However, not long after, the Romans managed to enter the city on a feast-day when the inhabitants were busy celebrating. The Roman general, Marcellus, ordered that Archimedes be taken alive, but when a Roman soldier found him, Archimedes was busy with a geometrical problem in the sand and told the man not to step on his circles. In response, the soldier killed him.

> **SEE ALSO:**
> ▶ The Third Dimension, page 54
> ▶ Algebraic Geometry, page 92

Archimedes demonstrated the mechanical advantage, or force-boosting power, of a pulley by showing that just one person could use a pulley to haul a ship across the water.

Equations

THE DISCOVERY OF IRRATIONAL NUMBERS LED TO A move toward geometrical methods being the preferred approach to mathematical problem-solving. However, this was only a temporary shift.

There are many limitations to geometrical methods. For instance, to multiply three values, the geometrical approach is to calculate the volume of a cuboid, the edges of which have lengths equal to those values. Although this method is clear and simple, it is also very limited. There is no geometrical way to multiply four variables, for example, because there are no four-dimensional objects.

Beyond geometry

However, slow progress was made. Sometime in the first century CE, a Greek mathematician

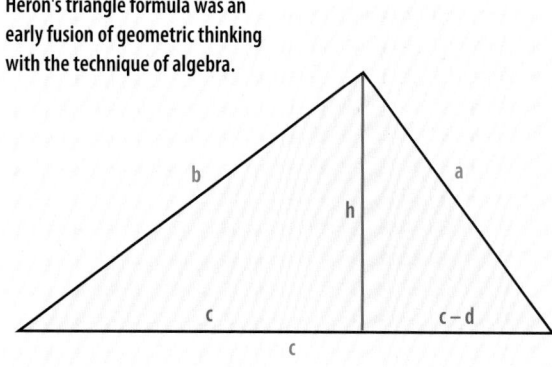

Heron's triangle formula was an early fusion of geometric thinking with the technique of algebra.

called Heron (sometimes known as Hero) wrote several engineering books, and in one of them he explained how to work out the area of a triangle based on the lengths of its sides. Although this sounds like a problem of geometry, actually it could not have been expressed properly by Greek geometry because it contains four variables—the three sides of the triangle a, b and c, and a fourth

Arithmetica, a book written by the Greek Diophantus about 1,900 years ago, is said to be the foundation of algebra.

variable, which we write as d, so d = (a + b + c)/2. Using modern notation, Heron's triangle formula is:

$$\text{area} = \sqrt{d(d-a)(d-b)(d-c)}$$

Heron, like all mathematicians before him, wrote out his books in words and numbers, which makes them very hard to follow today. In this sense, mathematics had hardly moved on at all since its earliest years.

A language of math

This all changed with the works of Diophantus, a Greek mathematician about whom we know almost nothing—not even the century in which

TYPES OF EQUATION

When an equals sign is used in math, the result is an equation, such as

$$x+1=4$$

An equation with a single unknown, like x+1 = 4, is known as **determinate**, because the value of x is determined by the equation. An **indeterminate equation** has more than one unknown, and the values of those unknowns cannot be determined by the equation on its own. An example is x+y = 4.

An equation that relates two or more variables is called a **formula**, such as v = d/t which relates the variables velocity (v), distance (d) and time (t).

So, $x+1=4$ is not a formula.

An equation with the same variables on both sides, and which is always true no matter what values the variables have, is called an **identity**.

$2(a+b) = 2a+2b$ is an example.

A collection of variables on its own, with no equals sign, such as 7 x 3, is not an equation, it is called an **expression**. A **function** is a relationship in which the input values always lead to just one output value. So, the function f(x)=x^2, if fed with the values x = 2 or x = −2, will output the value 4.

The symbol ≈ is sometimes used in equations. It means "approximately equal to." For example:
$$\pi \approx 3.14159.$$

he lived: He was born in either the 2nd or the 3rd century CE. We do have a very short biography, which is supposedly the inscription on his long-lost tomb. Unfortunately, it is in the form of a puzzle, so it's quite possible that the facts in it were changed to make it easier to solve:

> 'HERE LIES DIOPHANTUS, THE WONDER BEHOLD. THROUGH ART ALGEBRAIC, THE STONE TELLS HOW OLD: 'GOD GAVE HIM HIS BOYHOOD ONE-SIXTH OF HIS LIFE, ONE TWELFTH MORE AS YOUTH WHILE WHISKERS GREW RIFE; AND THEN YET ONE-SEVENTH ERE MARRIAGE BEGUN; IN FIVE YEARS THERE CAME A BOUNCING NEW SON. ALAS, THE DEAR CHILD OF MASTER AND SAGE, AFTER ATTAINING HALF THE MEASURE OF HIS FATHER'S LIFE CHILL FATE TOOK HIM. AFTER CONSOLING HIS FATE BY THE SCIENCE OF NUMBERS FOR FOUR YEARS, HE ENDED HIS LIFE.'

Can you figure out how old Diophantus was when he died? He was 84. Actually, this is a very unsuitable inscription for the man who did more than anyone else to get away from the idea that mathematical problems should be set out and solved in sentences. Diophantus's greatest contribution to algebra was to introduce symbolic notation, including an equals sign. As a result, it was at last possible to write equations and symbols for several positive and negative powers.

Seeds of future math

Reading Diophantus's book *Arithmetica* today, it is clear just how challenging it is for anyone to develop something new in math. It is full of hints of new kinds of math, but most are not developed by Diophantus. Since we know

Heron is also remembered for inventing the aeolipile, a primitive (but elaborately finished) form of steam engine.

How it works

Minus x Minus = ?
Although Diophantus could not accept negative numbers as the solutions of problems, he used them as steps in calculations, and that meant he had to grapple with the concept of multiplying them together. While it is fairly obvious that the product of a negative and a positive number is a negative number, whether the product of two negative numbers is positive or negative is not so obvious. Diophantus decided that the product is positive, but he did not prove it.

However, here is one way to do so. We want to prove that:

$$(-a)(-b) = ab$$

Let's define a number **x**, such that

$$x = ab + (-a)b + (-a)(-b)$$
(call this equation 1)

First, we factor out **b**

$$x = ab + (-a)b + (-a)(-b)$$

$$x = b(a + (-a)) + (-a)(-b)$$

$$x = b(0) + (-a)(-b)$$

$$x = (-a)(-b) \text{ (equation 2)}$$

Now we go back to equation 1, and this time we factor out **−a**

$$x = ab + (-a)(b + (-b))$$

$$x = ab + (-a)(0)$$

$$x = ab \text{ (equation 3)}$$

Equations 2 and 3 give two expressions that are both equal to **x**, which means that those expressions must equal each other.

That is:

$$(-a)(-b) = ab$$

Which is what we wanted to prove.

so little about him, it's impossible to say why. Perhaps he didn't appreciate the power of his ideas, maybe he was concerned that people might find them so far-fetched that they would reject the whole book, or perhaps he tried and failed to take them further. Diophantus only introduces a symbol for a single unknown. So, although he could write an equation corresponding to $x = 2$, he couldn't have written $x + y = 2$.

SOLVING EQUATIONS BY SUBSTITUTION

In order to solve simultaneous equations like these:

$x^2 + y^4 = 20$; $y^4 = 4x^2$

We can substitute the second equation into the first:

$x^2 + 4x^2 = 20$

$5x^2 = 20$

So

$x^2 = 4$ which means $x = 2$

And we can substitute that into the second equation:

$y^4 = 4x^2$

$y^4 = 4(2^2) = 16$

That gives:

$y = 2$

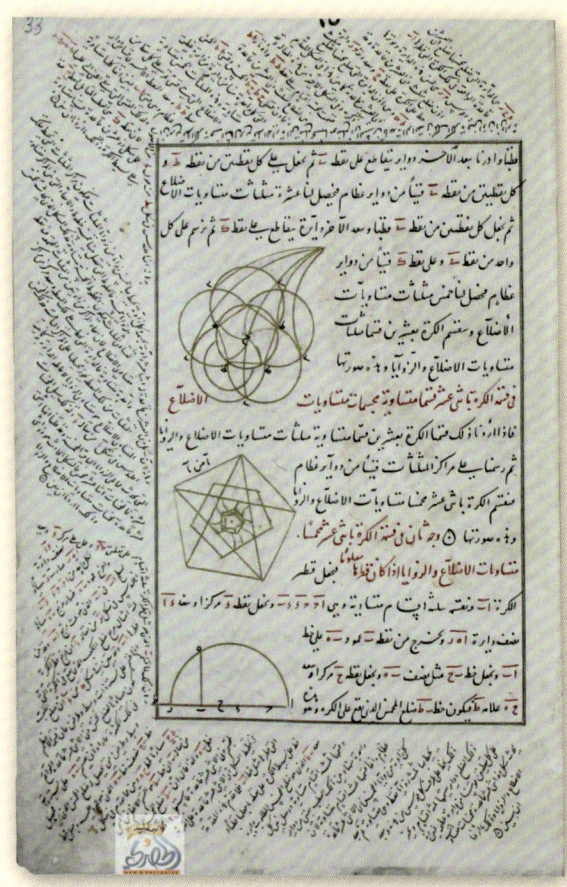

Arithmetica was translated from Greek into Arabic and was a major influence on Islamic mathematics around 1,000 years ago.

The strange thing is that problems involving multiple unknowns were very well known to him (and to earlier mathematicians). But when Diophantus discusses such problems, he has no option but to use the traditional approach of spelling them out in words, even though to us it seems obvious that, having invented a symbol for what we would call x, there was nothing to stop him going on to introduce y and z, too.

DIOPHANTINE MYSTERIES

A Diophantine equation only accepts solutions that are whole numbers. A great deal of modern mathematics deals with such equations, including research on prime numbers. Studying Diophantine equations might sound simple. For instance, it is much easier to study $a^x + b^y = c$ if we're only interested in integer values of x and y.

This means that we don't need to worry about bizarre things like $a^\pi + b^{-55.0098} = c$. But, this simplicity can be deceptive. For example, take $2^3 = 8$ and $3^2 = 9$. Are 8 and 9 the only consecutive whole numbers which are powers of other whole numbers? This question was asked by a Belgian mathematician called Eugène Catalan in 1844, but the answer ("Yes") was only proved in 2002, by Preda Mihăilescu.

Some Diophantine problems are still unsolved. For example, the whole-number solutions of Pythagoras's theorem are the Pythagorean triples like (3, 4, 5) and (5, 12, 13). What happens if we extend this idea to three dimensions, by making a cuboid whose faces are back-to-back right-angled triangles, the sides of which are whole numbers? (A cuboid is a three-dimensional shape with 6 rectangular sides.)

In that case, one side might look like this:

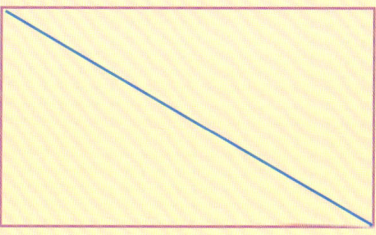

Eugène Catalan was still pondering the legacy of Diophantus 150 years ago.

And the whole thing would look like this:

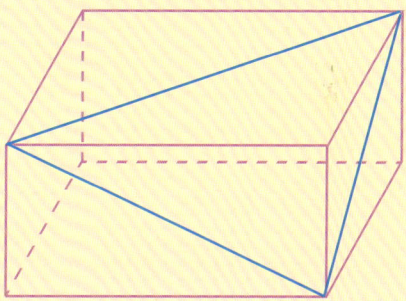

Finding such cuboids (called Euler bricks) is very difficult in itself. The smallest example has sides 44, 117, and 240 units long. But here's the real challenge: A perfect cuboid is a Euler brick in which the distance between opposite corners (called the body diagonal, as shown in green below) is also a whole number.

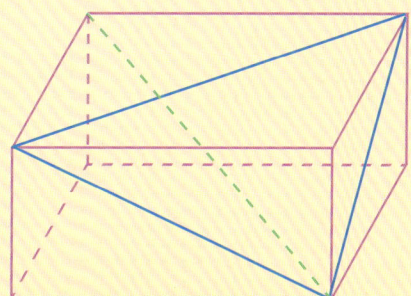

So far, no one has ever found one.

SUMS OF SQUARE NUMBERS

Diophantus's *Arithmetica* was translated from the original Greek into Latin (the international language of scholars in those days) in 1621, and in that version it influenced many mathematicians of the time (including Fermat, see page 99). The first was its translator, Claude Bachet. He spotted that, hidden in *Arithmetica*, was a startling idea, far ahead of its time: The conjecture that every whole number equals the sum of at most four square numbers (which are squares of numbers). For instance, $21 = 4^2 + 2^2 + 1^2$, and $127 = 11^2 + 2^2 + 1^2 + 1^2$.

This was eventually proved in 1770 by Joseph-Louis Lagrange, and is now known as Lagrange's four-square theorem. But that was not the end of the story. Also in 1770, Edward Waring suggested that there might be similar rules for all other powers. And so there are. In 1909 it was proved that every whole number can be written as the sum of at most 9 cubes, and that same year, David Hilbert proved that Waring was correct. For every whole number n there is another number m, such that any whole number can be written as the sum of at most m nth powers. But that proof didn't show how to find the values of m. It wasn't until 1986 that the value of m for n = 4 was found: 19. That is, every whole number equals the sum of at most 19 fourth-power numbers.

Joseph-Louis Lagrange was a leading figure in the French math community.

Glimpsing the future

Diophantus had a similar problem with negative numbers. He did tackle equations like $4 = 20 + 4x$, which took him very close to a full understanding of negative numbers. But, rather than concluding that $x = -4$, he says simply that the solution is nonsensical. He also shows in *Arithmetica* that multiplying numbers that are raised to powers is equivalent to adding those powers, such as

$$x^2 \times x^3 = x^{(2+3)}$$

So, for instance $100 \times 1{,}000 = 10{,}000$; this is the idea behind logarithms (see page 129), but he takes it no further.

Ahead of his time

Diophantus states some basic rules for dealing with equations, by either adding or subtracting terms from each side. So, for example, we can solve $x + 2 = 5$ by subtracting 2 from both sides, to give $x = 3$. Although he does not include the very powerful concept of substitution (see box, page 50) in his list, he does use it. From the perspective of modern mathematics, *Arithmetica* is an important mathematical milestone for another reason, too. It marks the introduction of the idea of a field in mathematics. A field consists of a type of number (such as rational numbers, real numbers, or complex numbers), the operations (addition, subtraction, multiplication, and division), and a set of rules (such as "$a + b = b + a$"). A lot of advanced algebra deals with fields and their relationships. Though ahead of his time in so many ways, Diophantus was behind them in another. Earlier mathematicians, going right back to Pythagoras, were comfortable with the idea of general solutions: one solution of $a^2 = b^2 + c^2$ is $a = 5$, $b = 4$ and $c = 3$, but there are many more than that. But, Diophantus gives no general solutions (at least, not in any of his surviving books). Like the Babylonians, he preferred just to give many examples of questions and answers, even though he did mention that some problems do have multiple solutions.

Times to come

Perhaps Diophantus's ideas were too far ahead of their time. For centuries, mathematicians took little notice of his discoveries, continuing to write out their problems in words and to ignore many of his insights. Arabian mathematicians did make some use of his ideas, but it was not until the late 16th century, in Europe, that Diophantus's breakthroughs in symbolic language and other areas were adopted and developed. Partly thanks to *Arithmetica*, mathematicians were able to develop whole new ways of doing math, no longer held back by the need to rely on clumsy sentences or fiddly diagrams. As a result, Diophantus is often called the Father of Algebra.

In 1970, the Russian Yuri Matiyasevich showed that, without working it out, it is impossible to know if a Diophantine equation has whole number solutions, thereby solving Hilbert's 10th problem (see below).

$$3x^2 - 2xy - y^2z - 7 = 0$$
$$x^2 + y^2 + 1 = 0$$

In 1900, German mathematician David Hilbert (seen here third from left in the front row) produced a list of 23 outstanding problems—a challenge to the mathematical community as the new century dawned. His tenth problem was to figure out if one could tell if a Diophantine equation had a whole number result, or not.

SEE ALSO:
▶ Cubics, page 64
▶ Differential Equations, page 116

The Third Dimension

IF ONE ROTATES A STRAIGHT OR CURVED LINE AROUND AN AXIS, then the resulting shape is called a solid of revolution. For example, if a triangle is rotated around its edge, it can generate a cone:

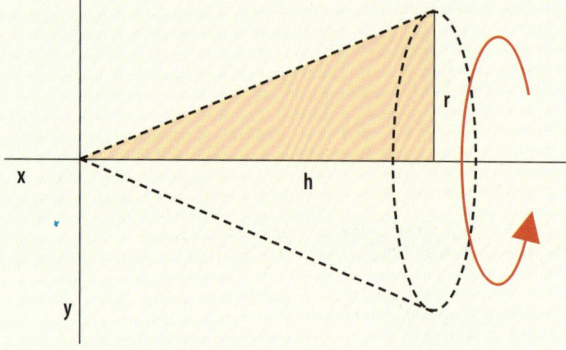

And a rotated rectangle can form a cylinder:

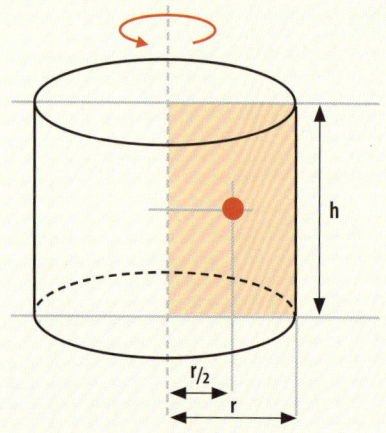

Trumpets, balls, bats, bottles, plates, eggs, jars... These are all solids that have an axis of symmetry.

Any solid shape with an axis of symmetry—a ball, jar, or bottle—can be thought of as a solid of revolution.

Pappus papers

Finding the volumes and surface areas of such shapes mathematically is an important application of integral calculus (see page 114). But, long before that was thought of, Pappus of Alexandria had shown that algebra can be used too, at least for simple shapes. We know almost nothing about Pappus, only the fact that he was a teacher in Alexandria, Egypt, and the name of his son was Hermodorus. Unlike many of the "facts" about other ancient Greeks, which come from biographies written by others, we can at least assume these statements are true, since they come from Pappus's own writings. We also know roughly when he lived, because he mentions a solar eclipse, which we can date as 320 CE, and he is quoted in a work written in about 411 CE.

Linking dimensions

Pappus's brilliant insight was to see that, since he could imagine a solid shape as being produced by rotating a flat shape, he might be able to do the same thing algebraically, by "revolving" the formula of the flat shape to produce the formula of the solid one. To calculate a volume of a revolution, we define the radius of the circle of revolution as the distance between the axis of the revolution and the center of the flat shape. For a simple shape like a rectangle, it's easy to see where the center is: halfway across, at ½r. So, the circle of revolution is $2\pi \frac{1}{2}r$, which is πr.

HOW FILLING IS MY DONUT?

A ring donut is known in math as a torus, and using Pappus's theorem to find the volume of one is simple if we "deconstuct" the shape. Cutting a cross section of this torus reveals a circle with a radius of r, and if this is revolved around a larger circle of radius R, then the torus would be generated.

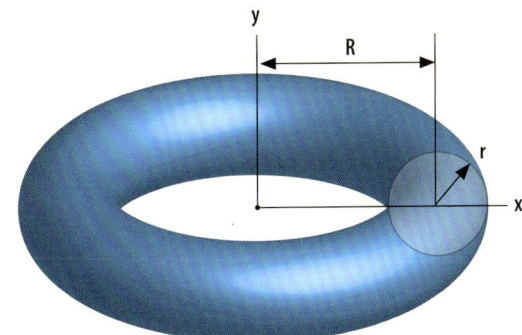

So, by Pappus's second theorem (see overleaf), the area of the torus is (area of revolved shape) × (length of path the shape is revolved around), which is:

$$\pi r^2 \times 2\pi R$$

Which is $2R\pi^2 r^2$

Then we multiply this by the area of the rectangle, which is hr, to give the volume of the cylinder formed by revolving a rectangle: $\pi r^2 h$.

Find the center

For a more complicated flat shape, like a semi-circle (which can be rotated to form a sphere), the position of the center is not so obvious. It is defined by the concept of a centroid, which was probably invented by Archimedes. The centroid of a shape is the point at which a panel of that shape will balance (perhaps better known as the center of gravity). So, if a cord is attached to the shape at that point, it will hang flat. For a semi-circular flat shape, the centroid is at a

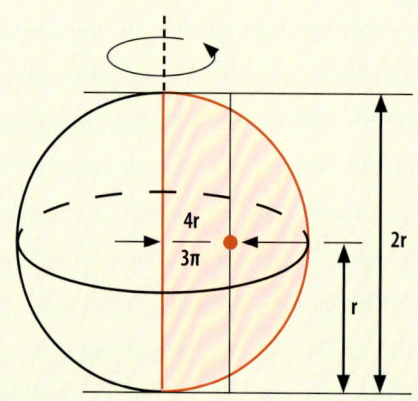

A sphere is a rotated semicircle.

distance of $4r/3\pi$ from the straight edge. So, the length of the circular path along which the centroid revolves is $2\pi(4r/3\pi)$, and the area of the semicircle is $\pi r^2/2$. The volume of revolution

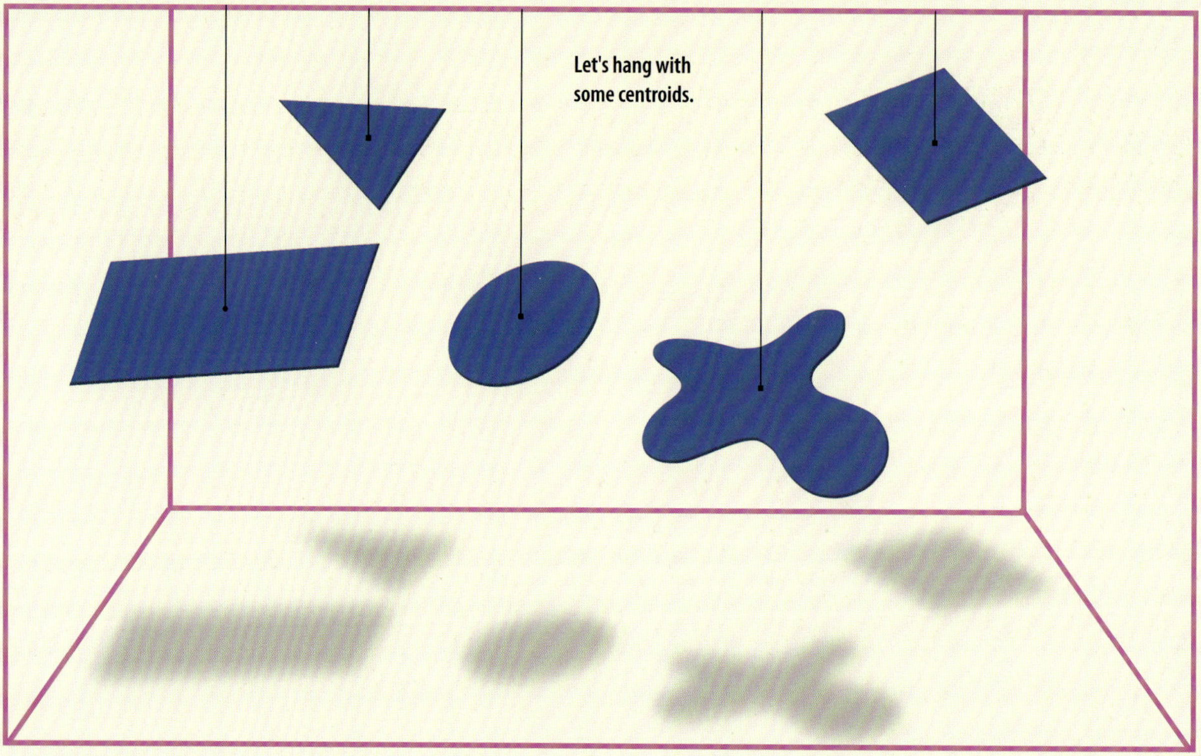

Let's hang with some centroids.

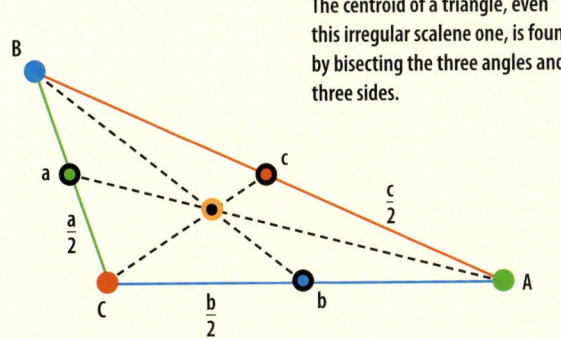

The centroid of a triangle, even this irregular scalene one, is found by bisecting the three angles and three sides.

is "path of centroid × area of flat shape" or $2\pi(4r/3\pi)\pi r^2/2$, which is $4/3\ \pi r^3$.

Second theorem first

This is all summed up by Pappus's second centroid theorem which, in modern terms, says:

The volume of a solid of revolution generated by the revolution of a shape about an external axis is equal to the product of the area of the shape and the distance moved by the shape's centroid.

The term "external" here means that the axis must not pass through the shape itself. This theorem can be used with quite complicated shapes (see box, page 55), and this idea of solids of revolution impacted on the art of the Renaissance, too (see box, page 59).

Kinds of surface

Pappus also showed how to determine the surface area of a volume of revolution. As usual in math, we need to be very precise about what we want to know. If we are only interested in the curved surface of the cylinder, as we would be if we wanted to know how much material we need to make a tube, then the area is given by the length of the line that is revolved (which is h, the height of the rectangle) multiplied by the length (circumference) of the circle of revolution. This circle has a radius that is again defined by a centroid, but in this case it is the centroid of the edge of the rotated shape (which is r), rather than the centroid of its area (½r). So, this time, the circle of revolution is $2\pi r$. Multiplying this by the length of the line that is revolved, h, gives our answer: $2\pi rh$.

First theorem

If the total area of a cylinder is required, the discs at its ends must also be included. Their areas are each πr^2. So, the total surface area is $2\pi rh + 2\pi r^2$. In modern terminology, areas of volumes

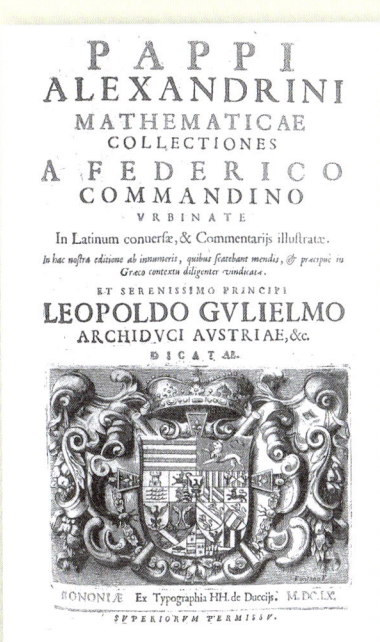

In order to encourage new mathematical research, Pappus wrote a twelve-book text called *Mathematical Collection*.

PAPPUS'S HEXAGON THEOREM

Take two rulers and paint three dots on each one, anywhere you like. Number the dots like this:

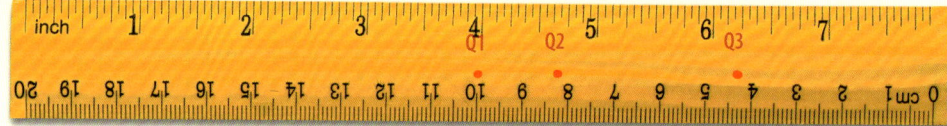

Drop the rulers. The positions of the dots on the floor are now random.
Draw straight lines from P1 to Q2 and Q3, P2 to Q1 and Q3, and P3 to Q1 and Q2.

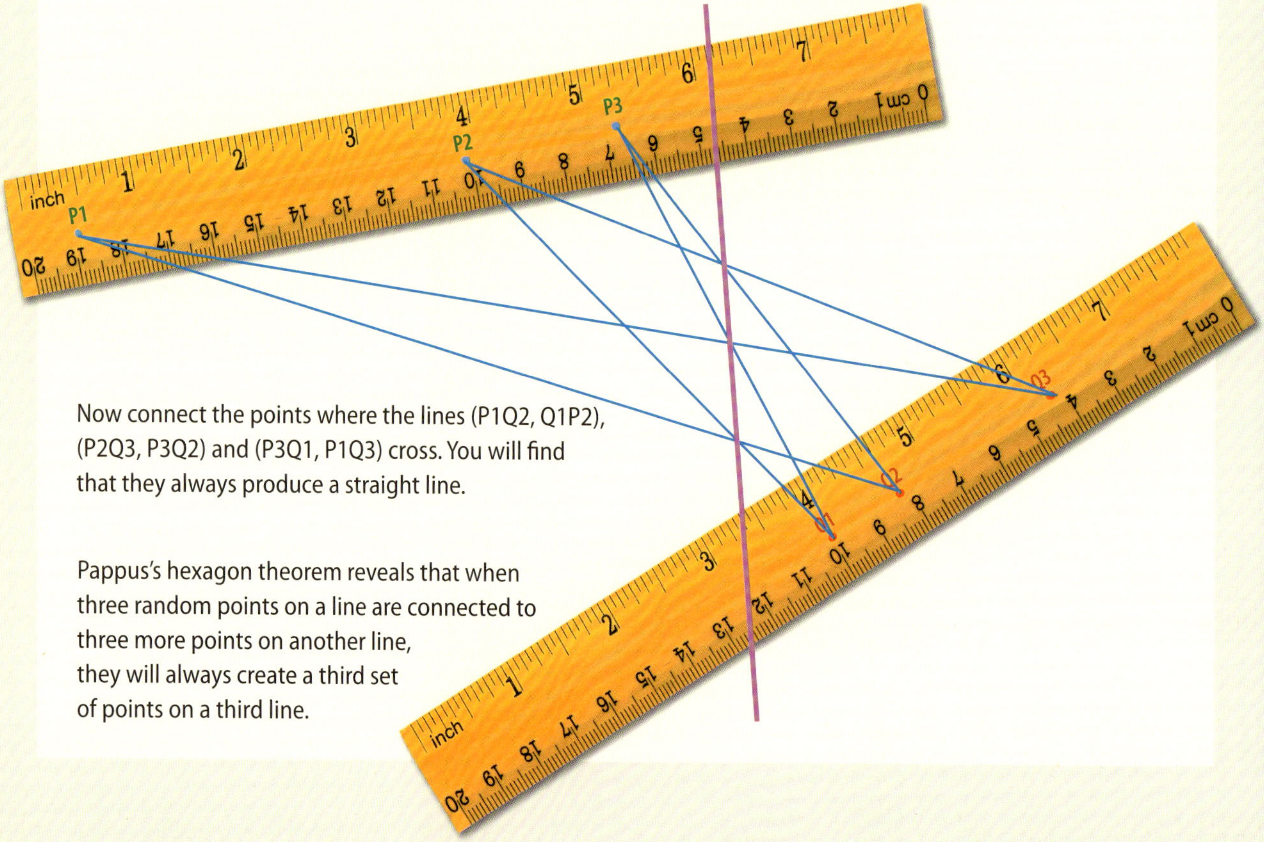

Now connect the points where the lines (P1Q2, Q1P2), (P2Q3, P3Q2) and (P3Q1, P1Q3) cross. You will find that they always produce a straight line.

Pappus's hexagon theorem reveals that when three random points on a line are connected to three more points on another line, they will always create a third set of points on a third line.

of revolution are calculated according to Pappus's first theorem:

The surface area of a surface of revolution generated by the revolution of a curve about an external axis is equal to the product of the length of the generating curve and the distance traveled by the curve's geometric centroid.

Among Pappus's other discoveries is the hexagon theorem, which is one of the best and earliest examples of a major theme of mathematics and science: to find order in apparently chaotic situations (see box, left).

Darkness falls

In Pappus's surviving writings, he comments on the lack of progress in mathematics in recent years. By this time, the Greek Empire had long fallen, and his city of Alexandria was part of the Roman Empire. The Romans seem to have had little interest in new mathematics themselves, and, once their empire too had crumbled away, the Western world became a poorly educated and uncivilized place, in the period sometimes called the Dark Ages.

> SEE ALSO:
> ▶ Finding the Maximum, page 86

THE ART OF REVOLUTION

The idea of rotating a flat shape to make a three-dimensional one was of great benefit to artists struggling to draw such shapes in perspective. This vase construction is by Paolo Uccello, an Italian artist of the mid-15th century.

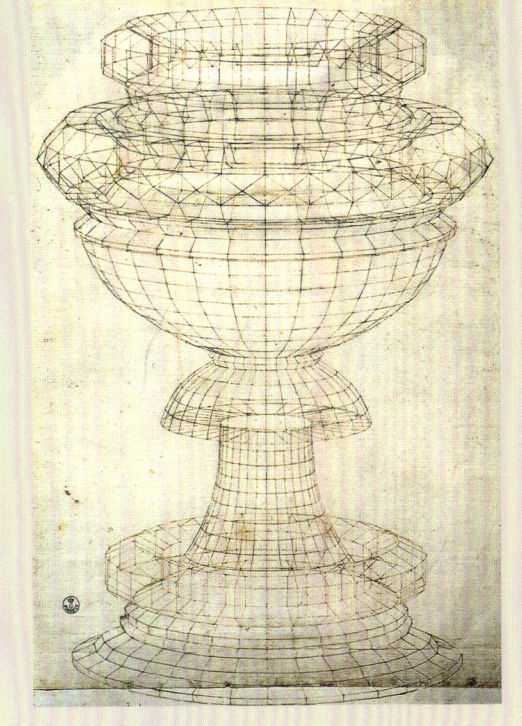

The wonders of ancient Alexandria faded after Pappus's time. By the 18th century, all that was left was Pompey's Pillar, which stands to this day.

Algebra Moves East

At the House of Wisdom in Baghdad, algebra became a globally used mathematical technique.

ALEXANDRIA, WHICH HAD BEGAN AS A SMALL EGYPTIAN TOWN before being taken over and enlarged by the Greeks and then the Romans, was the world's greatest mathematical center for many centuries.

But mathematics in Alexandria came to an end in the 5th century. In 415 CE, Hypatia, a female mathematician and teacher of astronomy, was murdered by a Christian mob during a period of religious intolerance and civil unrest. So, many scholars left the city for their own safety. In 641 CE, Alexandria was captured as part of the Islamic expansion. The mathematical knowledge of the Greeks that had been collected and developed there was studied by the Muslims who now controlled the city. Once Baghdad had been set up as capital of the Islamic empire in 762, it became the new center of knowledge.

Knowledge hub

In addition to the Greek science and mathematics that was studied there, the scholars of Baghdad could also draw on the cultures of Persia, India, and China (see box, overleaf). In about 800,

The Arabic title of Al-Khwarizmi's book gave us the word *algebra*.

How it works

Algorithms

Today, the word *algorithm* refers to an approach that is central to the way in which almost all calculations are done—by computer.

Unlike computers, people have the ability to carry out all sorts of math questions without the need to explain exactly how they do them. For instance, the mathematical method to fit a curve to points is quite complicated, but most of us can do a fair job freehand.

Similarly, what is the area of the shape on the right? One way to find out is to break the shape up into a couple of triangles, a rectangle and a semi-circle, and work out their areas before summing the total.

It would be very time-consuming to write down the precise instructions to do that for someone else to follow. In addition, there are other ways to break up the shape, so it would be almost impossible to write instructions on how to decide on the best collection of simpler shapes to select. But this is a task that a human can do in seconds without too much thought.

Humans, especially children, are excellent at learning techniques and trying them out in different situations, which is how we solve problems like this. But computers do not learn in this way. The only way to get a computer to solve a math problem is to tell it exactly what to do, step by step. But even that is of no use if every single step must be fed into the computer—it would be quicker to solve the problem yourself. The answer is an algorithm, or a mathematical process that can be defined exactly, and then repeated over and over again until the problem is solved.

MATHEMATICS IN INDIA AND CHINA

There was a thriving culture of mathematics in India from 1200 BCE, and much of this knowledge became known to the Islamic Empire. Centuries later, it was passed on, in turn, to Western Europe by Fibonacci (see page 66) and other Renaissance scholars. The decimal system and zero were especially important concepts. Chinese and Indian mathematicians developed algebra independently of the Greeks, but many of their methods (like solving numerical problems by drawing diagrams) and results (like Pythagoras's theorem) are almost identical.

Ancient Chinese mathematics was often concerned with astronomy and astrology.

an important academy called the House of Wisdom was built in Baghdad. There, the great works of science and mathematics could be studied and scholars could meet to discuss them, make their own discoveries, and pass them on to students. One of the greatest scholars to work there was Muhammad ibn Musa al-Khwarizmi, and his greatest work was the *Handbook of Calculation by Completion and Reduction*, the most important math book to be written since Diophantus's *Arithmetica*, six centuries earlier. In fact, our word "algebra" comes from its Arabic title, *Al-kitab al-mukhtasar fi hisab al-jabr wa'l-muqabala*. Our word "algorithm" (see box, page 61) comes from Algorismus, the Latinized version of Al-Khwarizmi's name.

More than a tool

In the book's title, "completion" means adding a term to both sides of an equation, and "reduction" means subtracting a term. These are the two main principles introduced by Diophantus (see page 47), and Al-Khwarizmi helped spread them throughout the Arab world. The book did not adopt the powerful mathematical symbolism invented by Diophantus. Instead, problems (including numbers) were written out in words, as the Babylonians and early Greek mathematicians had done. In fact, there was not much new material in the handbook, but it is of great importance because it influenced so many later mathematicians. Al-Khwarizmi also treated algebra as a subject in its own right, not just as a set of tricks and shortcuts to help with arithmetic. In addition, although the problems Al-Khwarizmi solved were mostly simpler than those tackled by the Babylonians or by Diophantus, the way he approached them was new. Instead of trying to find the solution to an equation directly, he first simplified the equation, identified what type it was, and then applied a method of solution based on its type. This classify-then-solve approach is the one all mathematicians use today.

SEE ALSO:
▶ The Rules of Algebra, page 82
▶ Abstract Algebra, page 158

SIEVE OF ERATOSTHENES

An early example of an algorithm is a method of finding prime numbers called the Sieve of Eratosthenes. It dates from around 200 BCE.

We start by writing as many numbers as we wish to "sieve." The definition of a prime number is "a number greater than 1 that can be divided only by itself and 1."

So, first we can cross out the "1" as, by definition, that is not prime. Any number that can be divided by 2 is not a prime, so now we go through each number in turn and remove it if it can be divided by 2, unless that number is 2 itself. (This "each number in turn" phrase is common in an algorithm, and is the kind of instruction that is very easy to program into a computer.) We repeat the process with numbers that can be divided by 3, other than 3 itself (again "repeat the process" is typical computer-speak).

We keep doing this with successive numbers ("successive" is another algorithmic term). And the final result is this, with all the primes "sieved out." In the kind of code that computer programmers write, this algorithm would look something like this:

```
Define array PRIMES[1 to 100]
Set all PRIMES values to 1
A[1] = 0
   For A = 2 to 100
      For B = 2 to 10
      Divide A by B
      IF there is not a remainder AND A does not equal B
THEN set PRIMES[A] to 0
While A <= 100
READ PRIMES[A]
IF PRIMES[A] = 1 THEN PRINT " '[A]' is prime"
END
```

(B goes to 10 because to find the primes in a sequence of n numbers, the sieve method only needs to continue until the numbers have been divided by $\sqrt{n}$.)

Cubics

Omar Khayyam wrote his mathematics book while in his twenties, before moving on to poetry in later life.

OMAR KHAYYAM IS MOST FAMOUS TODAY AS A POET. English translations of his work, *The Rubaiyat*, have been popular since they were first published in 1859. But he is also a key figure in the development of algebra, thanks to his *Treatise on Demonstration of Problems of Algebra*, which he wrote in about 1060.

Like Al-Khwarizmi, Omar Khayyam did not use the powerful symbolic language developed by Diophantus, instead writing out his problems in words. However, also like Al-Khwarizmi, his clear descriptions of mathematical techniques encouraged future progress. Khayyam's special interest was in cubic equations, which had first been studied by Archimedes and were also known to Diophantus. But Khayyam took a much more rigorous approach to them, classifying them into fourteen different types and describing how to solve them, including a method based on the intersections of parallelograms and circles which he may have invented (see box, opposite).

What is a number anyway?

Khayyam regarded mathematical techniques partly as a useful set of tools for the solution of practical problems, just as the Babylonians had. But he was also a philosopher of mathematics, and was interested in particular in the relation of mathematical ideas to real things. This is still probably the most important topic of the

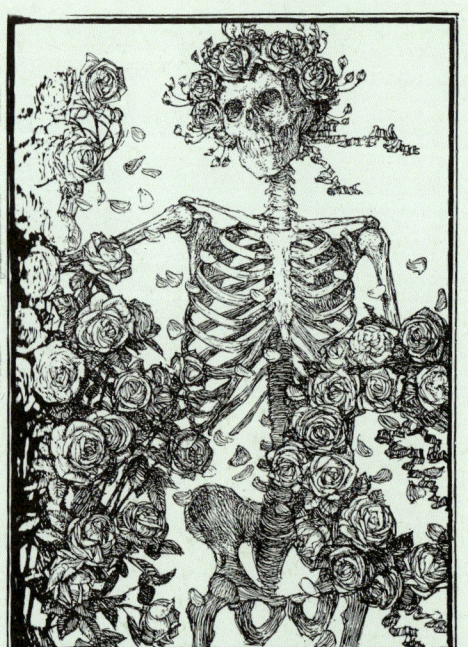

Left: A rather alarming illustration from the 1913 edition of the *Rubaiyat* of Omar Khayyam.

Right: A page from Khayyam's first but less celebrated book, this time on cubic equations.

philosophy of mathematics today. "Do numbers exist?" is a much trickier question than it might first appear. They do not seem to be real things in the sense that apples or books are, but on the other hand they are not completely imaginary things like dragons, either. "Two apples" is different from "three apples" in some entirely real sense. Perhaps we might say that numbers are like letters in that they are real things, but ones that we have invented. But again, the difference between "two apples" and "three apples" seems more like a discovery of a natural phenomenon than an invention—as if numbers are "out there" in a way that letters are not.

How it works

Solving cubic equations

Omar Khayyam's method can solve cubic equations of the form $x^3 + bx = c$ (although he would not have written them in this way). Here, it is used to solve $x^3 + 7x = 48$. A square with area $b = 7$ is drawn and a parabola drawn to pass through its top left and bottom right corners. A semicircle of diameter $48/7$ is drawn next to the square, and the solution (which we would call the x-value) is the horizontal distance from the right hand side of the square to the intersection point of the parabola and the circle. The answer is 3, as we can check by substituting $x = 3$ back into the equation: $3^3 + 7 \times 3 = 48$. Like all methods of the time, all that is needed to construct this diagram is a compass and a straight edge.

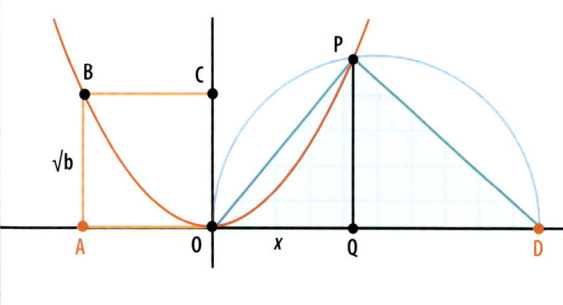

These questions don't only apply to numbers. The concept of cube roots sounds like it has been invented, but there is a real relationship between the volume of an actual cube and the length of one of its sides.

SEE ALSO:
▶ Unreal Numbers, page 74
▶ Pascal's Triangle, page 102

Sequences and Series

Fibonacci introduced the modern number system to Europe.

FOR MANY CENTURIES, the history of algebra was the story of great books. Diophantus's *Arithmetica*, Al-Khwarizmi's *Handbook of Calculation by Completion and Reduction*, and Omar Khayyam's *Treatise on Demonstration of Problems of Algebra* were each in their turn the most important books for students and researchers of mathematics.

The next great mathematician was Fibonacci, who also produced a great book, which was perhaps the most influential of all.

His *Liber Abaci* (The Book of Calculation) was published in 1202. It was this book that introduced to Western Europe the "Arabic" numeral system that is used all over the world today. The system it replaced was that of the Romans, which is extremely clumsy when used for calculations (see box, page 71).

African influence

Like Pythagoras, Fibonacci, whose actual name was Leonardo of Pisa, was a great mathematician partly because he was a great traveler. Though born in Italy, he was educated on the north coast of Africa, in Bugia (now Béjaïa in Algeria). Italy was a highly successful trading nation and some of that trade was carried out in Bugia by merchants from Pisa. Fibonacci's father was sent

٠١٢٣٤٥٦٧٨٩
0 1 2 3 4 5 6 7 8 9

there by the Pisan government to assist them, and his son joined him in order to learn to be a merchant. In the accountancy school in Bugia, Fibonacci may well have read the works of Al-Khwarizmi and Khayyam, and his work for his father involved travel to Syria, Egypt, France, and Greece. Such travels were not unusual for young Italian traders of his time, but what made Fibonacci different was his fascination with the mathematical systems he found on his trips. On his return to Pisa, he wrote *The Book of Calculation*, which included not only details of the Arabic number system (which actually originated in India) and its advantages, but also mathematical guidance for traders on subjects such as calculating interest.

Rabbit reproduction

The Book of Calculation also includes an account of the sequence of numbers that now bears Fibonacci's name. He arrived at the sequence by considering how the number of pairs of rabbits would grow over time, if left to themselves. If we assume that each pair produces one new pair each month, and that rabbits can breed from the age of one month old, then, starting from one newborn pair in January, there would be:

January: first pair

February: still just one pair

The famous sequence makes its first appearance in Fibonacci's book *Liber Abaci* (The Book of Calculation).

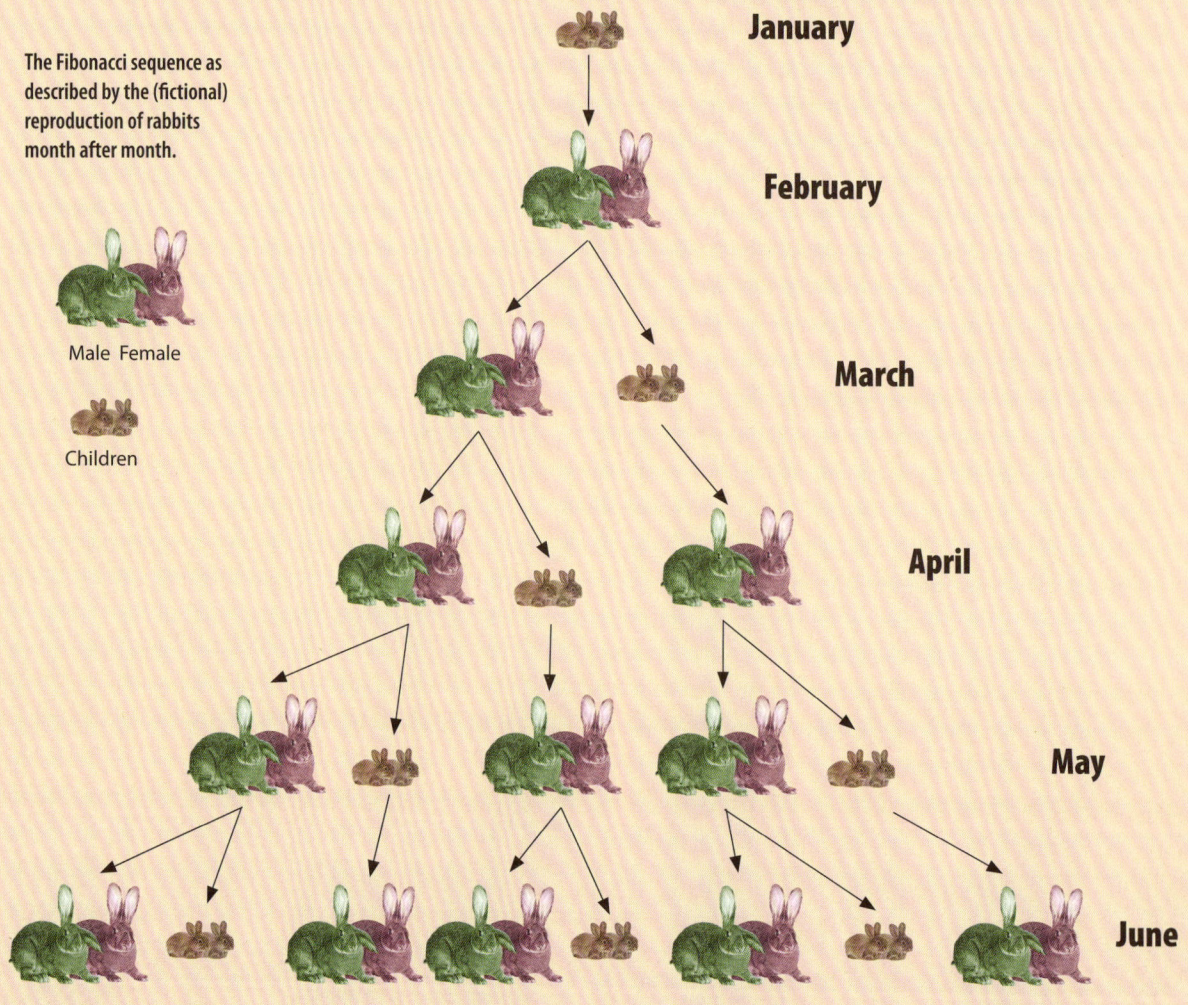

The Fibonacci sequence as described by the (fictional) reproduction of rabbits month after month.

Male Female

Children

March: first pair, and their children, Ron and Ruth = 2 pairs

April: first pair, plus their second pair of children (Rachel and Robert), together with Ron and Ruth = 3 pairs

May: Three pairs as above, plus a new pair of children for the original pair, and a pair of children for Ron and Ruth = 5 pairs

June: The above 5 pairs, plus a fourth pair for the original parents, a second pair for Ron and Ruth,

1, 1, 2, 3, 5, 8, 13, 21, 34, 55, 89, 144, 233,

Trees even branch in the Fibonacci sequence. The blue numbers show how many branches the tree divides into as it grows upward; the green numbers show how many branches are connected together at different levels.

and a first pair for Rachel and Robert = 8 pairs. It gets increasingly tricky to follow the sequence in words, but mathematically it is easy: Take the first two numbers, add them together (1 + 1 = 2), then add that new number to the previous one (1 + 2 = 3), and continue in that way (2 + 3 = 5, 5 + 3 = 8, 8 + 5 = 13 …). Each number (F_n) can be defined as $F_n = F_{(n-1)} + F_{(n-2)}$.

In nature

The Fibonacci sequence runs on forever: 1, 1, 2, 3, 5, 8, 13, 21, 34 … What is surprising is how often it turns up. Flowers have Fibonacci numbers of petals, sunflower seeds are arranged on the seed heads in curves containing Fibonacci numbers of seeds, and there are similar curves on pineapples and pinecones. Bee generations follow the sequence too: male bees have only a mother, while females have a father and a mother,

The spiral of scales on a some pinecones follows the Fibonacci sequence.

, 610, 987, 1597, 2584, 4181, 6765, 10946 …

which means that females have 2 parents, 3 grandparents, 5 great-grandparents, and so on.

However, despite the power of the Fibonacci sequence, it doesn't actually work for rabbits in real life! Rabbits have about six babies per litter, not two, and it is six months before they become fertile, rather than one. Fibonacci probably knew this, but simplified the situation to make it more mathematically interesting.

Other series

In math, a sequence is an ordered list of numbers: The integers, or whole numbers, form a sequence, for example. If the terms in a sequence are added together, the result is a series (see below), and the study of series is a major area of mathematics. Carl Friedrich Gauss (see page 130) was one of the greatest mathematicians, even while still at school. In 1787, his teacher tried to keep his class busy for a while by asking them

$$1 + 2 + 3 + 4 + 5 + 6 + 7 + 8 + 9 + 10$$
$$+ 11 + 12 + 13 + 14 + 15 + 16 + 17 + 18 + 19 + 20$$
$$+ 21 + 22 + 23 + 24 + 25 + 26 + 27 + 28 + 29 + 30$$
$$+ 31 + 32 + 33 + 34 + 35 + 36 + 37 + 38 + 39 + 40$$
$$+ 41 + 42 + 43 + 44 + 45 + 46 + 47 + 48 + 49 + 50$$
$$+ 51 + 52 + 53 + 54 + 55 + 56 + 57 + 58 + 59 + 60$$
$$+ 61 + 62 + 63 + 64 + 65 + 66 + 67 + 68 + 69 + 70$$
$$+ 71 + 72 + 73 + 74 + 75 + 76 + 77 + 78 + 79 + 80$$
$$+ 81 + 82 + 83 + 84 + 85 + 86 + 87 + 88 + 89 + 90$$
$$+ 91 + 92 + 93 + 94 + 95 + 96 + 97 + 98 + 99 + 100$$
$$= 5050$$

TROUBLE WITH ROMANS

The Roman numeral system is an awkward one to calculate with, partly because it has no zero, but mainly because it is not a positional system. It uses these symbols:

I (1)
V (5)
X (10)
L (50)
C (100)
D (500)
M (1,000)

Problems arise from the varying orders in which these letters are used. Sometimes pairs of numbers must be treated as sums **(XI = 10+1 = 11)**, but sometimes they must be treated as differences **(IX = 10–1 = 9)**.

When we need to add, subtract, multiply, or divide numbers and we can't work it out mentally, we usually write them down like this:

1979
+762

364
x 27

And then deal with each column in turn. We can do this because we know that the columns separate the number into ones, tens, hundreds, and thousands.

But in Roman numerals, **1979** looks like this:

Mathematicians argued for centuries about the merits of calculating with Roman or Arabic number systems.

MCMLXXIX, and the columns do not separate the number into smaller elements. The first **M** means one thousand, but we can only tell what the **C** means by looking ahead to the third character, which is an **M**. In this case we must subtract the **C** from it. The **LXX** means **50+10+10**, but the **I** can only be interpreted if we look to the number to the right of it, which is **X**. So, the **I** means **–1** in this case. In the end, we get **1000 – 100 + 1000 + 50 + 10 + 10 – 1 + 10**. Because we can't tell what a character means just from its position, calculations can't be done column by column, and become very confusing.

to add up all the numbers between 1 and 100. In less than a minute, ten-year-old Gauss had the answer: 5050. But great mathematicians are rarely especially good at mental arithmetic, and Gauss was no exception. He had reached the answer in record time by spotting that the answer he was looking for was the sum of a series: 1 + 2 + … + 100. And the sum of this series could be calculated in just three steps by using a simple formula.

Adding to 100

It's not clear whether Gauss had worked out this formula before or whether it was the teacher's question which led him to it, but this is how it works:

The series to be added has 100 members and runs from 1 up to 100. Imagine the reverse of this series. It also has 100 members but runs from 100 down to 1.

Carl Friedrich Gauss's genius was recognized at an early age.

Put the first series above the second. The first few terms will look like this:

1	2	3	4
100	99	98	97

If we add up each pair of numbers, it is soon clear that the sums are always the same:

1	2	3	4
+ 100	+ 99	+ 98	+ 97
= 101	= 101	= 101	= 101

We know that there are 100 terms, so the total of the two series must be

100 × 101 = 10100

And, to find the sum of just one of those series, we divide by 2, to get 5050.

An insight into how Gauss solved his school-room puzzle and entered the history books.

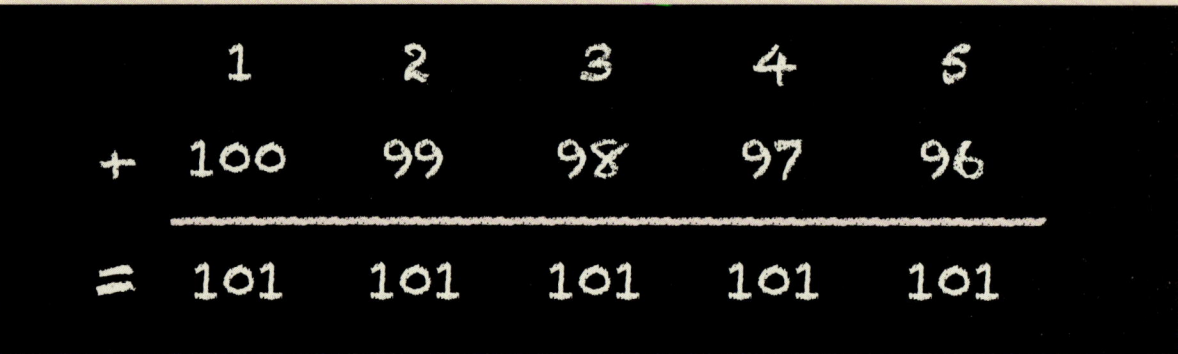

Of course, this would have worked however long the original series was, and we can state the general formula for the solution as

$1 + 2 + \ldots + n = n(n+1)/2$

Other series

Some series have surprising sums. For instance, the series

$$1 - \frac{1}{3} + \frac{1}{5} - \frac{1}{7} + \frac{1}{9} + \ldots$$

Sums to π/4, despite the fact that it seems to have nothing to do with circles.

The series

$$\frac{1}{2} + \frac{1}{4} + \frac{1}{8} + \frac{1}{16} + \ldots$$

Adds up to 1, which is fairly easy to visualize: Imagine a ruler 1 foot long. Color in 1/2 of it. Then color in half of what's left, which is 1/4 of the ruler. Keep coloring in half of what is left, which means coloring in 1/8 of the whole ruler, then 1/16 of it, and so on. Clearly, even though you would have to go on coloring forever to actually reach the end of the ruler, that is where you—and the series—are headed.

Convergent and divergent

This kind of series, which converges on an answer, is called a convergent one. Other series, such as the series $1 + 2 + 4 + 8 + 16 + \ldots$, obviously do not home in on a number, and are called divergent series. If we could add up all their terms, the answer would be infinitely large.

As is often the case in math though, it's risky to jump to conclusions about series. For instance, this series:

$$\frac{1}{2} + \frac{1}{3} + \frac{1}{4} + \frac{1}{5} + \frac{1}{6} + \ldots$$

may look like it's heading toward a number (2, perhaps?).

Actually, it is a divergent series. If we could add up all its terms, the answer would be infinitely large. This series is deceptive because its rate of growth is so slow. We would have to add up its first 12,367 terms just for the sum to reach 10, and even if we added the next 100 million terms to that, we still wouldn't have gotten to 20.

Coloring by halves: an infinite number of divisions adds up to a finite total.

SEE ALSO:
▶ Pascal's Triangle, page 102
▶ Calculus, page 110

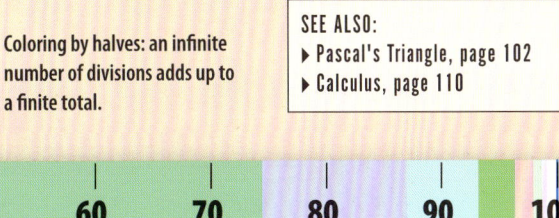

Unreal Numbers

WHAT ARE THE SOLUTIONS OF THE QUADRATIC EQUATION $5x^2 + 2x + 2 = 0$? As far as mathematicians were concerned, until the 15th century, there were no solutions for it. How could that be?

We can see why mathematicians believed this by using the general formula for the quadratic:

$$x = \frac{-b \pm \sqrt{b^2 - 4ac}}{2a}$$

with $a = 5$, $b = 2$, $c = 2$. The part inside the square root sign (sometimes called the discriminant) is then $2^2 - 4 \times 5 \times 2$, which

Niccolò Fontana Tartaglia is the first person known to have been able to solve all kinds of cubic equations. He made his name in math with his work on ballistics, or thrown bodies, as published in his book *Nova Scientia* (New Science).

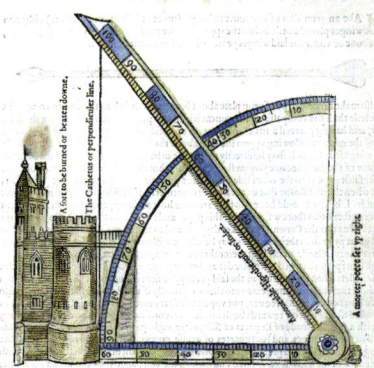

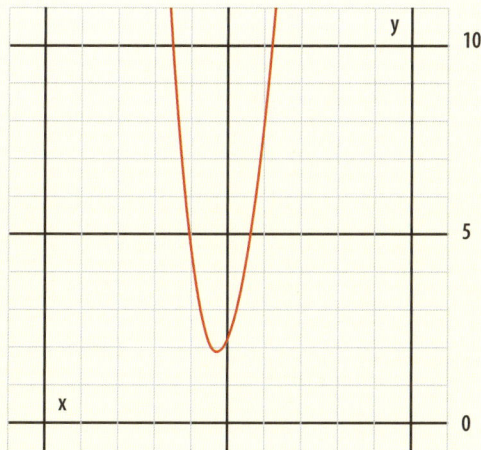

A line that never crosses the x-axis creates a problem.

is −36. And, −36 has no square root. There is no number that can be multiplied by itself to give −36. This can apparently be confirmed by drawing the graph of the equation (see above). The solutions of a quadratic are the values of y where the line crosses the x-axis; and since this line does not cross the axis, it has no solutions. Or so it seems.

Beyond geometry

Apart from the phenomenal successes of the ancient Greeks, mathematics had made very little substantial progress in the 4,000 years since algebra began in ancient Babylonia. It was only in the 15th century that this began to change. While the Greeks remained unsurpassed in terms of geometry (and some of their work was too complex even for the greatest of modern mathematicians to grasp), new discoveries in algebra began to be made.

In the 14th century, Arabic mathematical texts, including translated versions of Greek discoveries, began to appear in Western Europe. This was due largely to an increase in trade with the Near East, and it was the great trading nation Italy that benefited most. The new knowledge led many to make their own discoveries, initially and mainly in art, but gradually in science and math, too. This change is called the Renaissance, which means "rebirth."

Number battles

By the late 15th century, mathematics was becoming a popular pursuit in Italy, and there were even mathematical duels in which two mathematicians would compete to solve sets of problems by an agreed date. The loser's penalty was to provide a lavish dinner for as many of his opponents' friends as there had been problems to solve. Many of these duels concerned the greatest mathematical mystery of the day: the solution of cubic equations, that is, those of the form $ax^3 + bx^2 + cx + d = 0$.

While some examples had been solved by Archimedes, and a few others had succumbed to the efforts of mathematicians since, most cubic equations could not be solved. There were a number of things which made them especially hard to tackle. The first was that Diophantus's work was still being largely disregarded, so mathematical problems were still being set and solved (or not) in words. The other was that negative numbers were not considered an acceptable part of math.

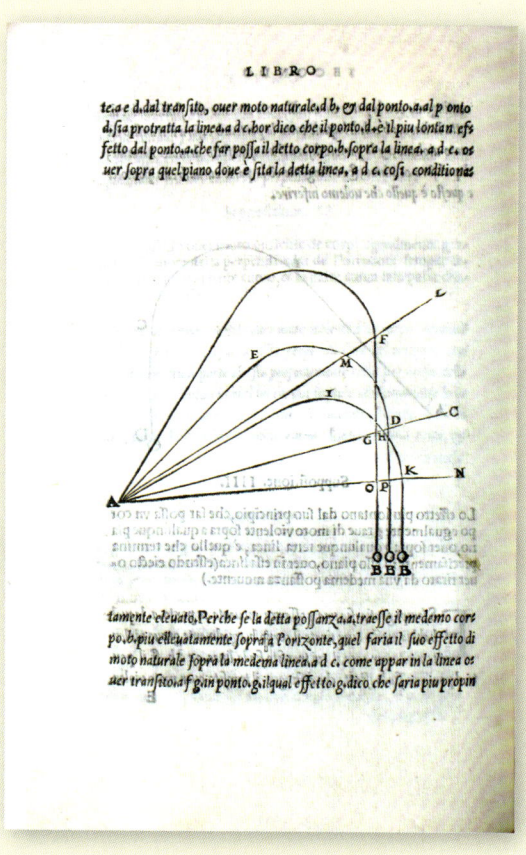

Tartaglia's early work on ballistics required an in-depth understanding of curves.

decided to use his secret knowledge to achieve fame and glory by triumphing over the great mathematicians of his day in mathematical duels.

Showdown

Fior's selected adversary was Niccolò Fontana Tartaglia, born in 1500 in Brescia. His name means "stutterer," thanks to the difficulties he had in speaking after suffering a throat injury during an attack on his town by French troops when he was a child. Tartaglia's father was a poorly paid courier, and Tartaglia was only half-

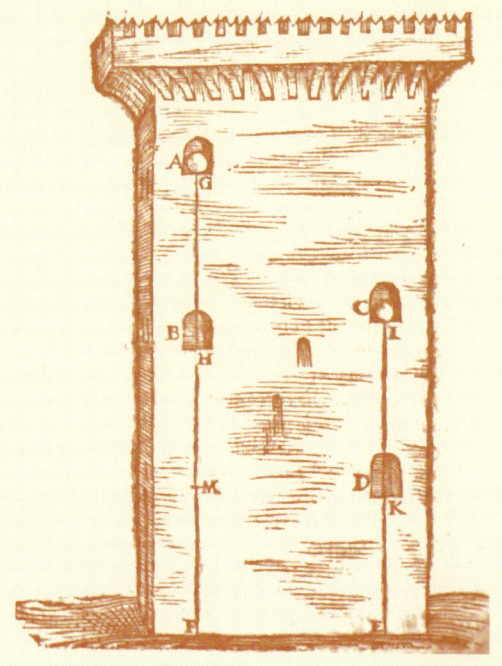

Secret solution

One particular duel that was to lead to the greatest breakthrough in mathematics for at least a thousand years began when a professor of mathematics in Bologna called Scipione dal Ferro discovered how to solve all cubic equations of this form: $x^3 = ax + b$. We don't know how he did it, and it's possible that he didn't. In those days, mathematical solutions were jealously guarded secrets. Ferro told only two people this particular secret, one of whom was his student, Antonio Fior. Once Ferro was safely dead, Fior

educated (literally—his family ran out of money to pay his school fees before his class reached the letter "L," so he left without knowing how to spell his name). However, he continued to learn on his own, taught himself math, and translated the works of Greek mathematicians into Italian. Because of this and because he also taught arithmetic, he was sufficiently well known for Fior to challenge him. Tartaglia's task was to solve 30 problems, all of them versions of the above cubic.

The pretender wins

As the cubic was thought to be unsolvable, Tartaglia apparently did not take Fior very seriously, since he didn't try to solve the problems for several weeks. However, according to the story, not long before the deadline of the duel (12 February 1535) he heard about Fior's secret. Panicked, he worked desperately on the equation and, just eight days short of the deadline, he found a method of solution. After that, the problems were easy to work out; he finished them all within two hours. That's the account that has come down to us, though it does sound very strange—why would Tartaglia have accepted a challenge he thought no one could solve? And why did he suddenly panic when hearing that Fior could solve cubics? It seems more likely that he knew the trick all along.

The deadline came and went without Fior solving a single one of the problems that Tartaglia had set for him, so it was Fior who had to provide dinner for thirty—or would have done, had Tartaglia not generously let him off. For Tartaglia, the fame was enough.

Enter Cardano

Because it is easy to check whether the suggested solutions of cubic equations are correct, simply by substituting them in place of the x-values, there was no need for Tartaglia to reveal his method. The outcome of the duel quickly became well known, and many people asked Tartaglia how he had solved the cubics, but he refused all requests until he met Gerolamo (or Hieronymous) Cardano, who was a very strange person indeed.

Cardano believed that he had supernatural powers. He claimed that he could determine the state of a person's internal organs using "second sight," and that his very presence was enough

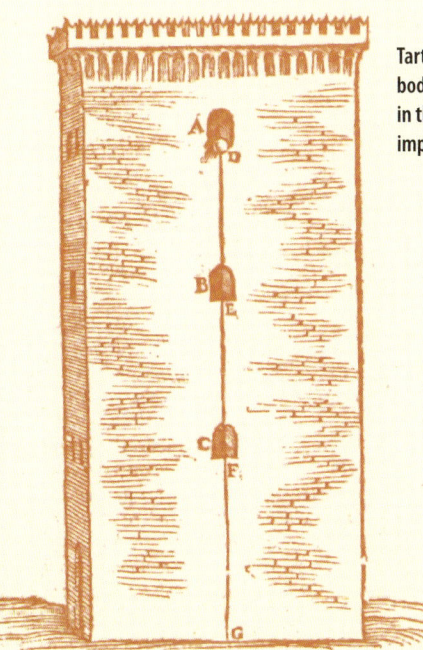

Tartaglia's work on falling bodies was the leader in the field until it was improved upon by Galileo.

to stop wounds bleeding. He was, in fact, a very successful doctor, so much so that he was admitted into the College of Physicians in Milan, despite their usual rule that children of unmarried parents, like Cardano, could not be accepted.

Cardano also claimed that he could predict the way dice could fall. It's therefore rather surprising that, according to his own account, he regretted the forty years he spent gambling at chess—and it's even more surprising that when he turned to gambling at dice, he hated it even more. However, his experience allowed him to become the first mathematician to analyze the role of chance in games, inventing probability theory. And he must have been fairly good at gambling, since he managed to make enough money to live.

Difficult character

Cardano also managed to upset a great many people. This was partly because he annoyed the Catholic Church by casting Jesus's horoscope and partly because of what he described as "the habit, which I persist in, of preferring to say above all things which I know to be displeasing to the ears of my hearers. I am aware of this, yet I keep it up wilfully." Cardano managed to control this habit when it came to Tartaglia, to whom he sent many polite requests for the secret method, but with no success until he promised to introduce him to the powerful governor of Lombardy. Tartaglia still refused to disclose the secret, but was very keen to

The title page of *Artis Magnae*, the most famous work of Gerolamo (or Hieronymous) Cardano.

Right: In 1554, Gerolamo Cardano made this drawing of a system for raising water using a series of screws—all powered by the stream's current.

meet the governor, so he accepted the invitation to Cardano's home. No one knows quite what happened next, but Tartaglia left almost at once, without having met the governor—and having given Cardano the secret method (in the form of a poem, oddly enough), on the express understanding that Cardano was to keep it totally secret and reveal it to no one whatsoever.

The Great Art

Since the secret was highly complex and written in words (some of which were no doubt chosen because they rhymed nicely, rather than being particularly helpful) it took years for Cardano to disentangle it, and not long after he did he published it in his book *Artis Magnae* or *Ars Magnae* (The Great Art), which was breaking his promise to Tartaglia of absolute secrecy in a rather big way. Tartaglia was furious and invited Cardano to another mathematical duel. Annoyingly, Cardano didn't turn up but sent his secretary instead. Even more infuriatingly, the secretary turned out to be a mathematical genius, and Tartaglia lost.

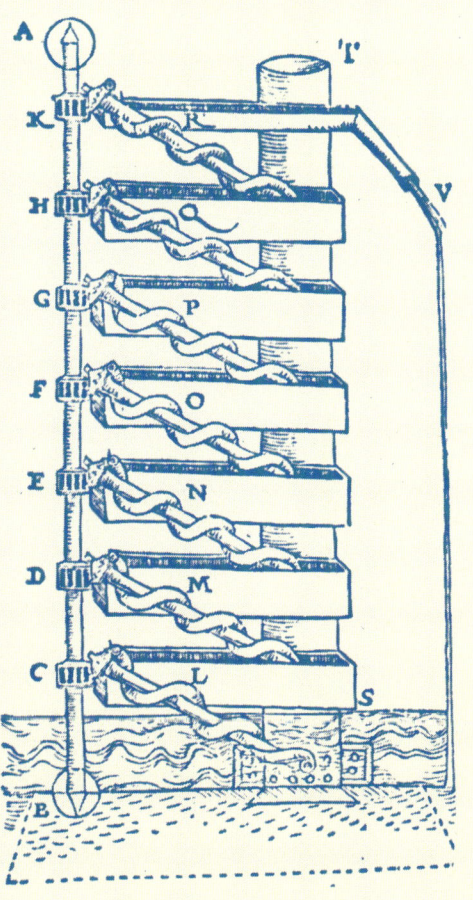

Negative roots

In working on Tartaglia's formula, Cardano found that sometimes it would lead to negative numbers, or to square roots of negative numbers. Like all previous mathematicians, he did not accept that either of these concepts were worth considering, referring to negatives as "purely false." He did, nevertheless, explore them briefly in *The Great Art*. For instance, he tried to solve the equations

$$x + y = 10$$
$$xy = 40$$

The solution he finds is:
$$x = 5, y = \pm\sqrt{(-15)}$$

Imaginary numbers

Today, we would represent that $\sqrt{(-15)}$ as $\sqrt{15}i$, which is about $3.873i$ and is called an imaginary number, the meaning of i just being shorthand for "the square root of -1," and a very powerful and useful concept. However, Cardano was not impressed. No sooner had he produced the world's first example of an imaginary number as a solution of an equation than he rejected it, saying that it is "as subtle as it is useless." After writing *The Great Art* in 1545, Cardano traveled to Scotland, where he became court physician to King Edward VI, then just 15. Cardano cast his horoscope and predicted several things to look forward to in the long life that lay ahead of the king. However, when Edward died just a few months later, Cardano hurriedly returned home.

Cardano predicted that his own death would take place on 21 September 1576. According to some, to show once and for all that he really did have special powers, Cardano killed himself on that date.

Finding reality

While Cardano had not thought imaginary numbers worth considering, they were explored thoroughly by Rafael Bombelli, an engineer from Bologna, Italy, who was in his twenties when *The Great Art* was published in 1545. Five years later, he had written a draft of his own great work, *Algebra*. Unfortunately, Bombelli was such a perfectionist that no parts of *Algebra* were published until 1572, and not all of it until 1929. The history of math might have been very different if it had all been published earlier, since it is revolutionary in several ways.

Using equations

For one thing, Bombelli followed Diophantus in using symbols. Bombelli was, in fact, the first person ever to write a quadratic as an equation—which seems amazing considering that they had been one of the main interests of mathematicians for at least 4,000 years.

But perhaps even more important, it is the first book in which imaginary numbers were taken seriously since they were first described by Diophantus over a thousand years earlier. Bombelli laid out the rules for their manipulation (see How it Works, right).

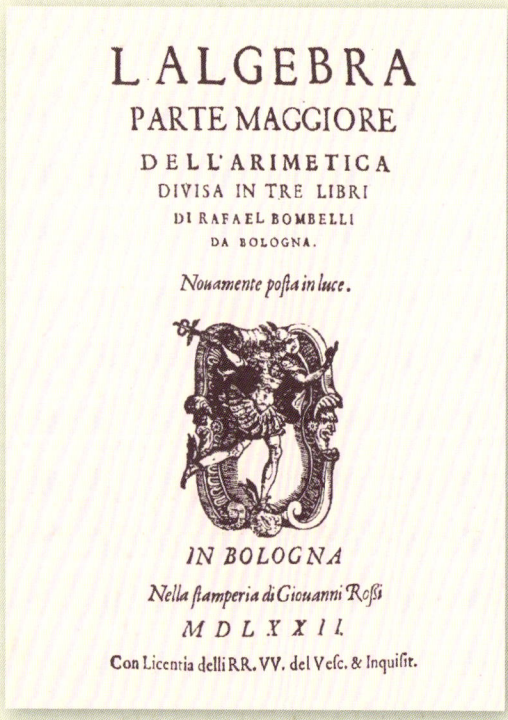

The title page of Rafael Bombelli's 1572 work *Algebra*, which extended the use of complex numbers.

SEE ALSO:
▶ Quaternions, page 148
▶ Abstract Algebra, page 158

How it works

Imaginary and complex numbers

Bombelli and today's mathematicians rarely use imaginary numbers in isolation, but refer instead to complex numbers, which have both a real and an imaginary part, like this: $a + bi$. So, one example of a complex number is $4 + 3i$.

Often, the two parts of complex numbers are now regarded as coordinates which can be graphed on what is called the complex plane. $4 + 3i$ looks like:

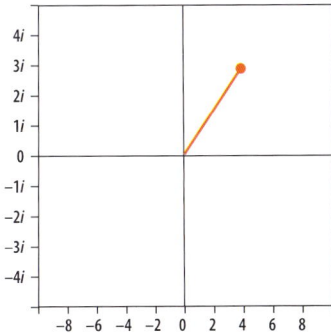

This is called an Argand diagram, after Jean-Robert Argand who came up with the idea in 1806.

The key point to bear in mind when manipulating complex numbers is that real and imaginary numbers are like apples and pears. You can do sums with either, but not with both together at the same time.

Adding and subtracting complex numbers is fairly simple, for instance:

$(4 - 3i) - (2 - 5i) = 4 - 3i - 2 + 5i = 2 + 2i$

Multiplication is trickier. To multiply $(a + bi)$ by $(c + di)$, we carry out 4 multiplications in turn, writing down the result of each, to get $ac + adi + cbi + bd(i)^2$. Since $(i)^2 = -1$, this is the same as $ac + adi + cbi - bd$. Therefore:

$(4 - 3i) \times (2 - 5i) = 8 - 20i - 6i - 15 = -7 - 26i$

Complex division is ... complex.

Given $\dfrac{a + bi}{c + di}$

We start by calculating something called the complex conjugate of the denominator, which is simply the same number with the central sign reversed, that is, $c - di$.

Then, we multiply top and bottom by that complex conjugate

$$\dfrac{(a + bi)(c - di)}{(c + di)(c - di)}$$

To give $\dfrac{ac - adi + cbi + bd}{c^2 + d^2}$

So $(4 - 3i) \div (2 - 5i)$

Is $\dfrac{8 - 20i - 6i + 15}{2^2 + 5^2}$

Which is $\dfrac{23 - 26i}{29}$

Finally, we separate this into real and imaginary parts, so the final answer is

$$\dfrac{23}{29} + \dfrac{-26}{29}i$$

The Rules of Algebra

TODAY, MATHEMATICAL RESEARCH IS ALL ABOUT PROVING THINGS clearly using algebraic symbols, and sharing the results widely by means of clear and concise papers (released online to begin with).

François Viète is the father of algebraic symbology.

In the 16th century, things were very different. Mathematicians worked mainly with words, numbers, and diagrams, and would often keep their methods secret for months or even years. When they did publish their discoveries, they were usually contained in lengthy books that would be filled with numerous examples of solved problems written out in words and numbers, not very different to the Babylonian mathematical clay tablets. Apart from being long-winded, these lists of examples only prove that the theorem or technique works in some situations and do not show why it works. To do that, a proof is needed, but 16th-century algebra books rarely contained them.

A new language

This was very different to the way that the ancient Greeks had worked: their mathematical discoveries were rigorously proved. The reason for this is that the Greeks had the powerful mathematical language of geometry at their disposal. Being unable to improve on the Greeks in terms of geometry, 16th-century mathematicians tried to explore algebra, but without an effective language to do so.

The man who changed all this was François Viète, born in 1540 in France and trained as a lawyer. In 1564, Viète took a job as tutor to the daughter of a noblewoman, a connection which led to him being made a member of the government of Brittany by Charles IX in 1573. This was despite that fact that Viète was a member of the Huguenots, a group of French Protestants whom the king, a Catholic, had ordered to be massacred the previous year. But Viète was successful in his new post, and, when Charles was succeeded by Henry III, he became one of the new king's legal advisors in Paris. However, as anti-Huguenot feeling in the city grew, his position became increasingly difficult until, in 1584, he was sent away.

Code-breaking

In 1589, King Henry summoned Viète to Tours, where he had set up his new parliament. That same year, Henry was assassinated, but the new king, Henry IV, not only welcomed Viète, but gave him a new job: To decode messages that had been intercepted on their way to King Philip II of Spain. Viète managed to do this (see box, page 85), and Philip, unable to believe that anyone could break his code, complained to the Pope that France was using black magic against Spain, and claimed that Viète was the devil.

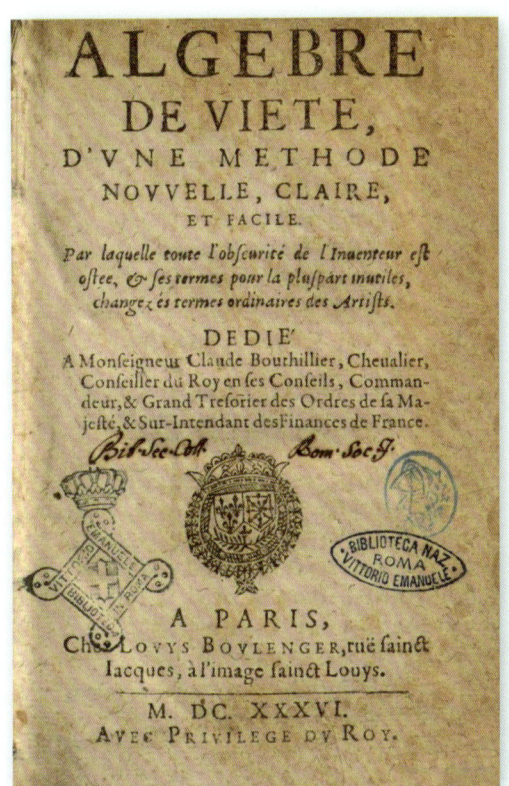

Viète's most famous book did not have the most imaginative title.

Variables and constants

Viète continued to serve Henry IV until 1602, and died the following year. But, his enforced break when he was banished from the Catholic court for a few years was the best thing that could have happened to him. With no duties to perform, he was able to spend all his time working on the subject he had always loved—mathematics. He had a clear objective, and a very bold one, too: To invent a new kind of math, one that would combine the power of Greek geometry with the much wider range of problems that algebra could tackle. The key to this new approach would be a mathematical language of symbols, with defined rules. Although variables had been represented by symbols before, Viète was the first to use them for constants as well.

Nowadays, we are used to expressions like $ax^2 + bx + c = 0$, which represents a quadratic equation. The use of symbols for the coefficients (numbers that multiply variables), a, b, and c, as well as for the variable (x) means that this single expression stands for all possible quadratic equations. This is a very powerful idea because it

Viète's life's work was completed under the patronage of King Henry IV of France, who defeated the previous king in 1594.

means that, if we can solve this general equation, we can solve every quadratic that exists. And in fact, we do know the solution of that equation:

$$x = (-b \pm \sqrt{(b^2 - 4ac)})/2a$$

Anything and everything

With general formulas like this, math became much easier to do (just plug in your chosen a, b, and c values, and do some sums). It also became much easier to explain. For instance, this is how Cardano's (see page 77) explanation of how to solve a quadratic equation begins: "Cube the third part of the number of unknowns to which you add the square of half the number of the equation, and take the square root of the whole which you will use, in the one case adding the half of the number which you just multiplied by itself …" Not only is this very confusing in itself, it means you spend all your time trying to work out how to actually do the math, not thinking about why it works or how you might improve it. Using words also obscures the power of the formula itself. In the above formula, for example, the "$b^2 - 4ac$" part is known as the discriminant, and it's important because it tells you how many real roots (or solutions) the quadratic equation has (or to put it another way, how many times a graph of the quadratic will cross the x-axis). If $b^2 > 4ac$ there are two different real roots. If $b^2 = 4ac$ there are two equal real roots, and if $b^2 < 4ac$ there are none. This sort of insight is very tricky without the general form.

Also, the reason that algebra books written without symbolic language rarely contain any proofs of the theorems they discuss, is that most types of proof (see page 18) can only be used with symbolic language. It's not surprising, bearing all this in mind, that algebra had made such little progress since the Babylonians, or that mathematicians spent nearly 4,000 years trying to sort out quadratic equations. And it's also not surprising that, after Viète published his ideas in *The Analytic Art* in 1591, progress in algebra became so much more rapid.

> **SEE ALSO:**
> ▶ The Fundamental Theorem of Algebra, page 130

DECIPHERMENT

The coded messages that Viète unscrambled have long been lost, and Viète gave no details of how he worked out their meanings, but they were probably examples of substitution ciphers, in which every letter of the alphabet is replaced by a number or other symbol.

Viète's method seems to have been based on what we would call letter-frequency tables. By counting the number of times each letter of the alphabet is used in a long piece of text, it's found that some letters occur much more often than others. In English, the most popular letters are e, t, a, o, i, and s, and in Spanish (the language of the letters Viète deciphered), they are e, a, o, s, n, and r. However, there are two major problems with this method. For one thing, the frequencies vary a lot from text to text, so much so that it is only really the first four or five most common letters that one can be sure of. And, a big sample of the ciphered text is needed before anything useful can be done. The first letter Viète decoded contained about 500 characters, which is quite short for this method to work. For the seven commonest letters he would have expected to find about 69 e's, 42 a's, 36 o's, 34 s's, 32 n's, and 26 r's. But the "about" here makes a big difference: The actual figures probably varied by at least 15 percent (so there were probably between 58 and 80 e's). So, while Viète could probably be confident in finding the characters for e and a, the frequencies of o, s, and n are so similar that he could not have been sure which was which.

Viète's breakthrough came when he guessed that large numbers (which were not ciphered) in the text referred to sums of money, which meant that the characters following them probably meant "ducats" (an international currency). But it was still a very arduous process of trial and error to work out what other characters might mean. As he said himself: "One must note all the sorts of figures, whether ciphers or jargon, and count how many times they occur, then note all the sorts of figures that precede or which follow and compare the most frequent in order to discover the same words, and the same meanings. Don't spare either labor or paper."

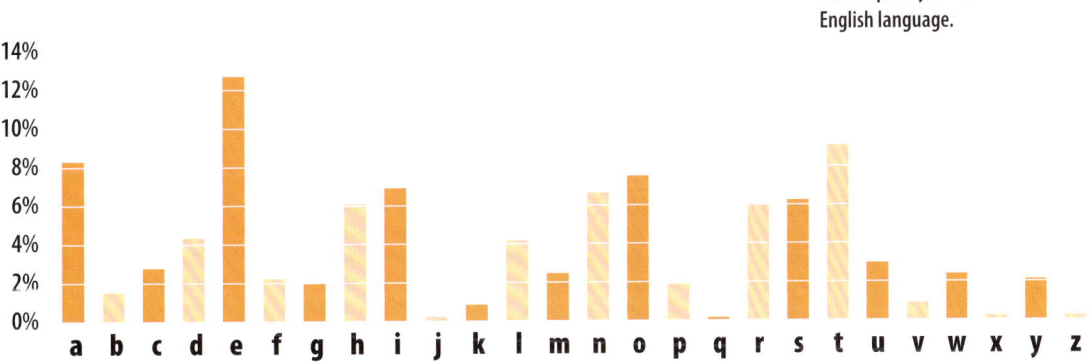

The frequency of letters in the English language.

Finding the Maximum

IN 1613, JOHANNES KEPLER, AN ASTRONOMER AND MATHEMATICIAN famous for finding the mathematical laws that describe the motion of the planets, and who was soon to marry for the second time, went to the wine merchant to order wine for the wedding.

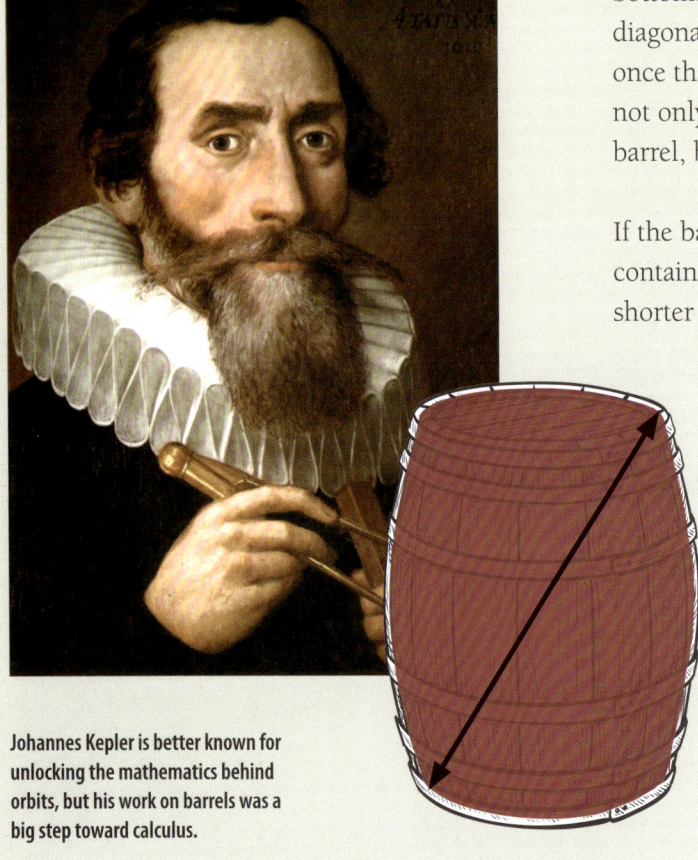

Johannes Kepler is better known for unlocking the mathematics behind orbits, but his work on barrels was a big step toward calculus.

To work out the bill, the merchant measured the amount of wine in a full barrel by inserting a measuring rod diagonally across it, from top to bottom. The price was set by the length of the diagonal. Kepler was unimpressed, realizing at once that the result of the measurement would not only depend on the volume of wine in the barrel, but on the barrel's shape, too.

If the barrel was very tall and narrow, it would contain much less wine than a somewhat fatter, shorter barrel with the same diagonal length. The ideal barrel for the wine buyer would be the one with the shortest diagonal for its volume.

Which barrel?

To find out whether he was being cheated, Kepler calculated the volumes of many wine barrels with the same diagonal, but different dimensions. (In order to make the calculations simpler, he assumed that they were all cylinders.) He was looking for the barrel with the largest volume, as that

MAXIMIZING THE VOLUME OF A CYLINDER BY GRAPHING

The blue line on this graph shows the volumes (in cubic inches, although any units would do) of cylinders with a diagonal length of 50 in, plotted against the ratio of their height to their width. The maximum volume occurs in a cylinder where the ratio is approximately 0.7. That is, when the height is 28.7 in and the diameter is 41.0 in (28.7/41.0≈0.7). This is indicated by the red line. The green line is the tangent to the curve which passes through the point where $x = 0.7$. The fact that the tangent is horizontal shows that its slope is zero; in other words, the volume is not changing here. It was Kepler who realized that, as the maximum is approached, the volume change falls toward zero. Today, plotting the graph of such a function is an obvious move, but the idea of graphing a function was only developed after Kepler's time. The most accurate way to find this maximum volume is to differentiate the function and, given that the differential gives the rate of change, we would then find the point at which the differential falls to zero.

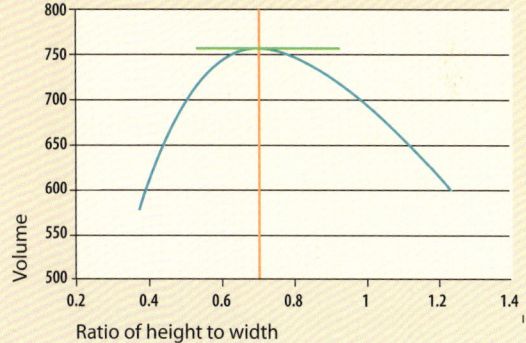

would have held the most wine for the price he paid. Today, this is easy to work out from a graph (see box, above), but graphs had not been invented in Kepler's time.

A good deal?

Kepler was doing by hand what calculus would one day do in one step: finding the maximum value of a function. He was pleased to find that, for a series of cylindrical wine barrels with the same diagonal but different diameters and heights, the one with the largest volume had a similar height and diameter to the barrel the merchant had used. So he had got about the maximum amount of wine possible for the price.

Shape and size

Kepler also found something interesting. For barrels much thinner, or much stubbier, than the best one, just a small change in shape made a big difference in volume. But, as the volume approached its maximum possible value, changes in the height or diameter made little difference. This was good news because it meant that, as long as the barrel the wine merchant had used was fairly similar in shape to the best possible one, it would contain almost the same amount of wine. Although this seems a trivial point, it was to be essential to the way calculus worked. Calculus is often used to find maximum (or minimum) values of functions, and it does this

MAXIMIZING THE VOLUME OF A CYLINDER BY CALCULUS

The formula for the volume V of a cylinder is
$$V = h\pi r^2$$

In that equation, the volume V is a function of two variables, the height h and the radius r. Can we make our task easier by making V a function of just one variable?

If we look at the cylinder in cross section lengthwise, it is like a rectangle with height h and width 2r.

Our diagonal divides this rectangle into two right-angled triangles, so we can express its length by Pythagoras's theorem:

$$d^2 = h^2 + (2r)^2$$

This means

$$r^2 = \frac{1}{4}(d^2 - h^2)$$

So, we can get rid of the r^2 in the equation for V by substitution, to give

$$V = \frac{h\pi}{4}(d^2 - h^2)$$

$$= \frac{\pi}{4}(hd^2 - h^3)$$

Now, how does the volume change if we change h? In other words, what is the rate of change of V with respect to a change in h? To find out, we differentiate the equation for V with respect to h (see page 111 for more details about differentiation).

$$\frac{dV}{dh} = \frac{\pi}{4}(d^2 - 3h^2)$$

As Kepler (almost) spotted, at a maximum or minimum, the rate of change of a function falls to zero.

So, when we maximize V, dV/(dh) must be zero, that is:

$$\frac{\pi}{4}(d^2 - 3h^2) = 0$$

So

$$(d^2 = 3h^2)$$

Therefore

$$\frac{h}{d} = \sqrt{\frac{1}{3}} \approx 0.6934$$

Which is the value we can see on the graph in the box on page 87.

by finding the point at which the change in the function falls to zero (see box, above).

This was not the end of the story. Although Kepler was now happy with his wine, he realized that, for practical purposes, the few simple shapes that the Greeks had studied were insufficient. He had been forced to assume that the wine barrel was a cylinder, rather than using the formula for the correct shape. However, a formula for barrel

Finding the Maximum > 89

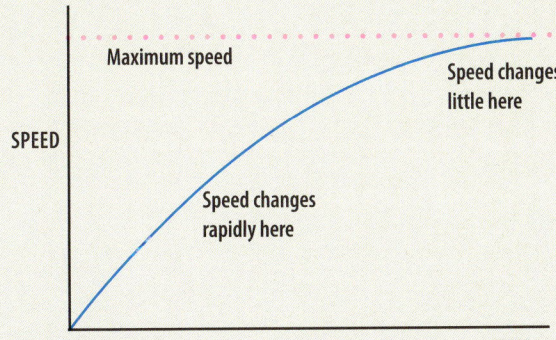

A graph showing the increasing speed of a car, with the driver accelerating as quickly as possible. To begin with, speed increases rapidly (the slope is steep) but as the top speed is approached, the rate of increase slows (the slope flattens). This is what Kepler found from his study of wine barrels, and it is this effect which calculus uses to identify maximum and minimum values.

almost straight to strongly curved. So, while a cylinder can be defined by just two numbers (height and width), a barrel needs many more numbers to define it.

Building a shape

Kepler's solution was basically the same method that Archimedes had used to find the area under a curve (see page 41)—to get closer and closer to the actual answer by adding together smaller and smaller bits. Kepler adapted that method by dividing up the total volume of a barrel into many disc-shaped slices, calculating their areas, and adding them all together. The smaller and more numerous the discs became, the closer the sum of their volumes came to the volume of the barrel.

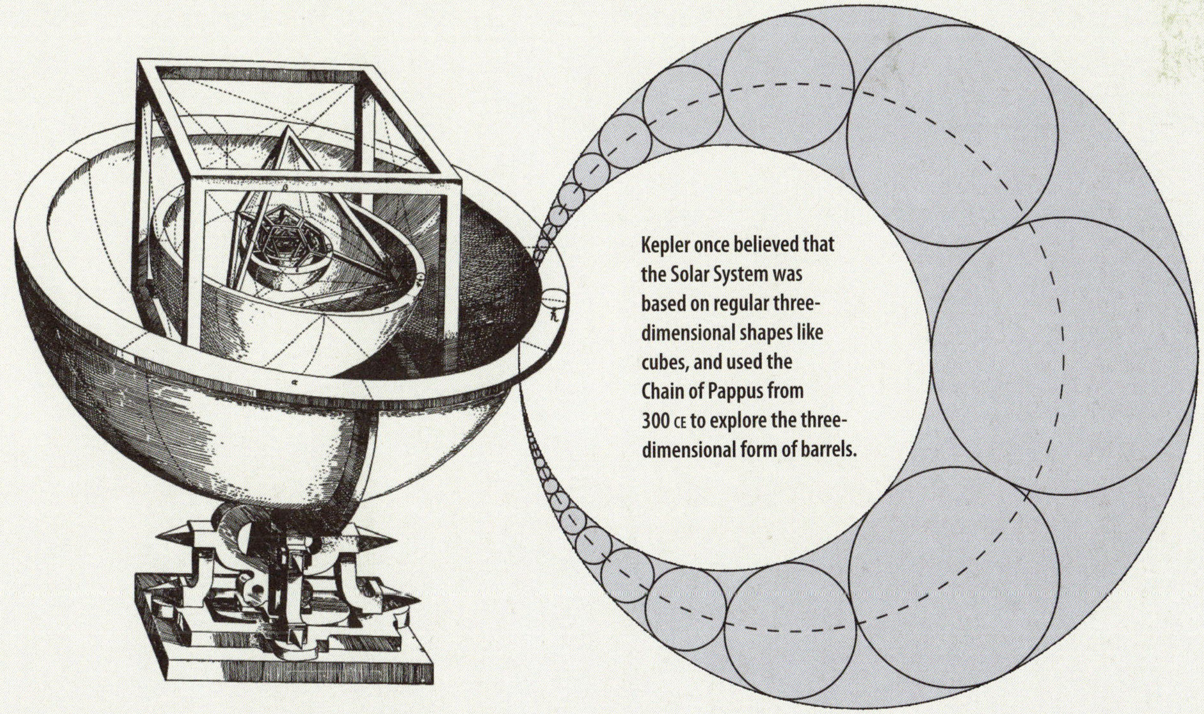

shapes would be very complicated because their shapes vary such a lot, with sides ranging from

Kepler once believed that the Solar System was based on regular three-dimensional shapes like cubes, and used the Chain of Pappus from 300 CE to explore the three-dimensional form of barrels.

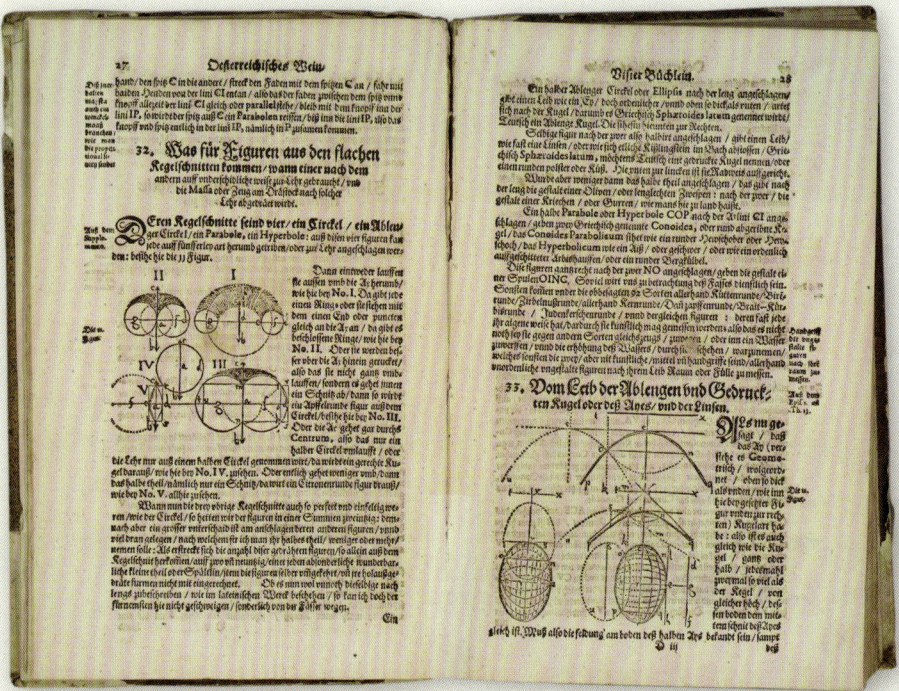

Kepler's 1615 work *New Solid Geometry of Wine Barrels* drew on ancient math of solid shapes.

A new look at an old approach

Using this method and also Pappus's theorem to calculate volumes of revolution (Pappus was probably the first person to calculate the volume of a torus, see page 55), Kepler, a very thorough and enthusiastic man, ended up with the volumes of 92 solid figures, which he described in his book *New Solid Geometry of Wine Barrels*, published in 1615.

Kepler's approach is found throughout mathematics. For example, the converging series on page 73, which get closer and closer to a final value with each new term. However, this was not Kepler's only non-astronomical mathematical work. An earlier study was to keep mathematicians busy for over four centuries.

Close packing

One snowy day in 1611, Kepler was crossing the famous Charles Bridge across the Vltava river in Prague when a snowflake landed on his coat. He was too poor that winter to buy a New Year's present for his friend Johannes von Wackenfels, and he imagined what a nice gift a snowflake would make. He did the next best thing, and wrote Johannes an essay on why all snowflakes have six corners. He decided that the reason must be that snowflakes formed when "frozen globules" packed themselves tightly together, and that they could achieve the closest packing by forming hexagons.

The idea that hexagonal packing was the closest had been suggested to Kepler by

A classification of the shapes of snowflakes made by Israel Perkins Warren in 1863.

was a better arrangement. Harriot decided that a hexagonal frame would allow the maximum number of cannonballs to be stacked into the smallest space. Kepler agreed, saying that hexagonal packing "will be the tightest possible, so that in no other arrangement could more (spherical) pellets be stuffed into the same container." This became known as Kepler's conjecture, but it was only proved in 2014.

British navigator Thomas Harriot's ideas about cannonballs had a wide-ranging impact on mathematics, chemistry, and crystallography.

Thomas Harriot, an English mathematician who was employed as a navigator by Sir Walter Raleigh, the explorer who set up the first English-speaking colony in North America and helped to introduce tobacco and potatoes to Europe. On Raleigh's ships, cannonballs were stacked into square or triangular frames on deck, but Raleigh wondered if there

SEE ALSO:
▶ Calculus, page 110
▶ e, page 122

Algebraic Geometry

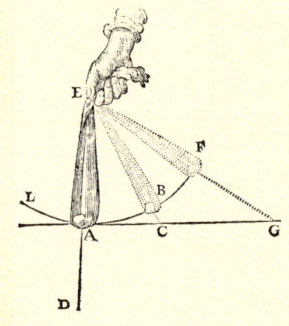

René Descartes imagined the motion of objects in terms of straight lines and curves described by algebra. His system also turned math into lines.

ALTHOUGH DIAGRAMS HAVE BEEN USED TO SOLVE MATHEMATICAL PROBLEMS FOR THOUSANDS OF YEARS, the idea of drawing a graph of an equation does not seem to have occurred to anyone until the 16th century.

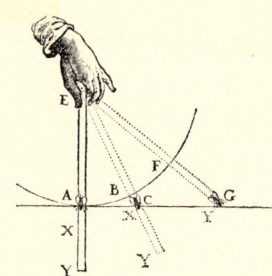

Perhaps this is partly because the idea of a variable wasn't yet clear. Many mathematicians still worked with examples of calculations—for example: "A man works thirty hours a week and is paid two pennies an hour. He gets 60 pennies a week or 240 pennies a month …") rather than using equations, such as: "The total pay equals hourly rate of 2 pennies an hour multiplied by number of hours worked."

Cash for math

Faced with the equation, we might draw a series of lines on a graph to show the man's income for different numbers of hours and different rates of pay, and use it to answer questions about his total income in different situations. But, working from the example, it's difficult to see what kind of graph we could draw, or why we would want to.

This began to change with Viète's introduction of a symbolic language (page 82). The person who tidied up the small difficulties with that system in order to introduce the algebraic language we still use today was also the mathematician who grasped the power of graphs to solve algebraic problems: René Descartes.

Proving himself

Descartes worked in many fields of study, including astronomy, biology, and physics, but it is a philosopher that he is best remembered today. In search of certainty about the world, he imagined that an all-powerful "evil genius" existed, who could make people see, hear, and feel anything he liked, and fool them into believing in things that didn't exist. Descartes concluded that, no matter how powerful the evil genius was, there was one thing he could not deceive Descartes about: Descartes' own existence. "I think, therefore I am," as he put it.

From this absolutely certain foundation, Descartes believed that he could be certain about many other things, too, including mathematical truths and even basic laws of physics. In fact, Descartes' most important mathematical work (*The Geometry*) is tucked away at the end of one of his major philosophical texts, *Discourse on*

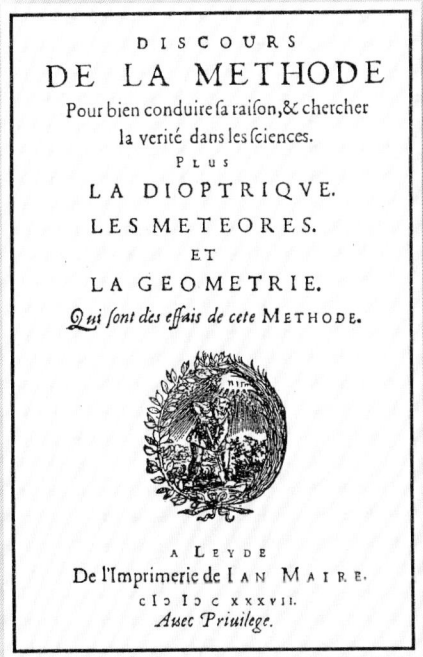

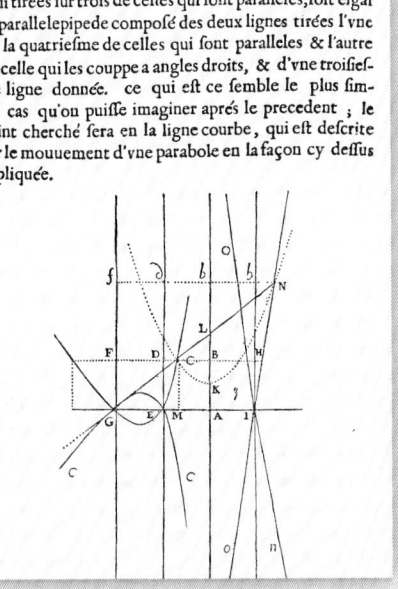

The concept of coordinates that related algebra with geometry was almost a footnote in René Descartes' masterwork.

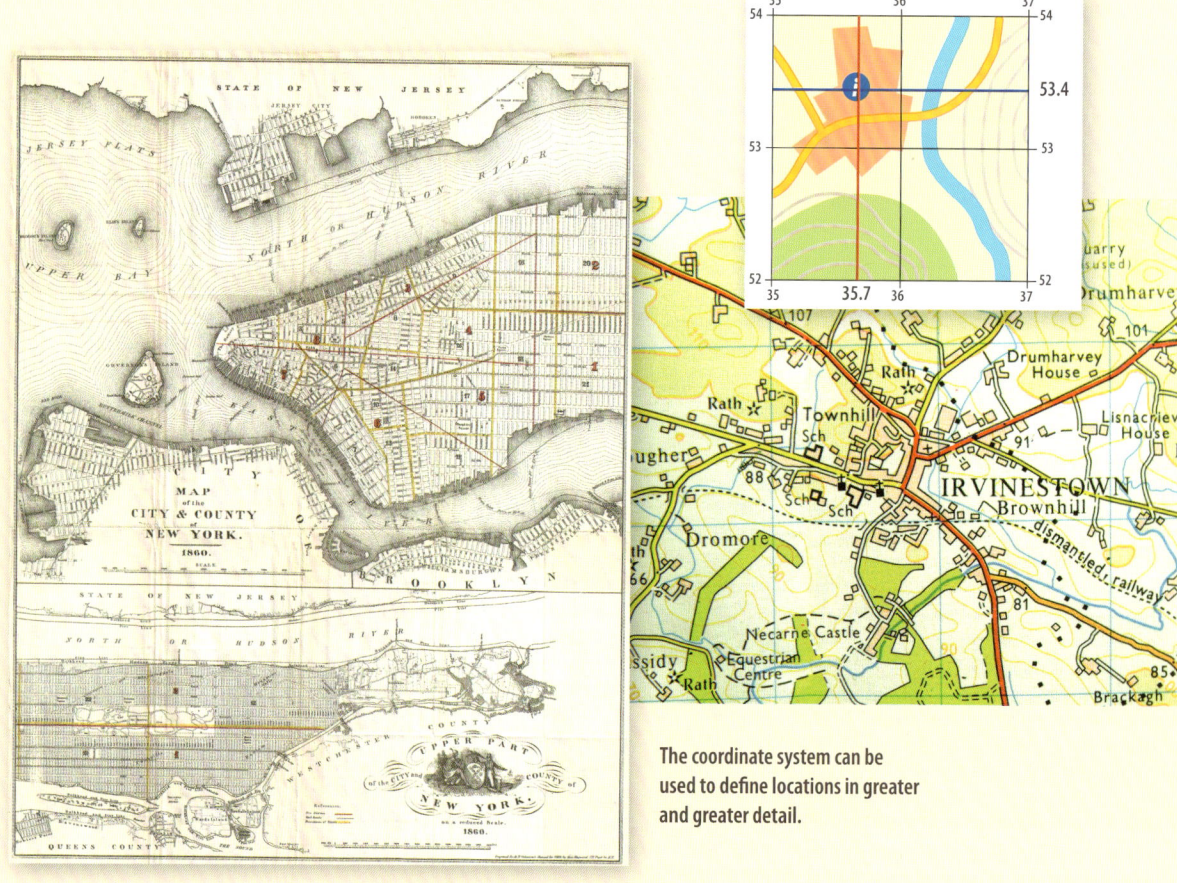

The coordinate system can be used to define locations in greater and greater detail.

the Method of Rightly Conducting the Reason, and Seeking Truth in the Sciences.

Places in space

The key to Descartes' mathematical breakthrough is the concept of a coordinate: A set of numbers which defines a point in space. Coordinates work in a similar way to locations in parts of New York City, where the roads are laid out in a grid pattern with Avenues running north to south and Streets running west to east. The location of the Port Authority Bus Terminal is at the intersection of 42nd Street and 8th Avenue; we might say that its coordinates are (42, 8). A more sophisticated version is used on geographical maps: a map reference (or grid reference) gives the location of any point on the map. So, in the top map, the Tourist Information Centre (the white letter "i") is approximately at map reference 357534 (the first three numbers indicate the "easting" and the last three are the "northing").

What has this got to do with mathematics? Well, we can map many things other than just landscapes, and coordinates can represent more than just eastings and northings. A line graph

THE LIFE OF DESCARTES

Descartes was a strange mixture. From his school days, he found he did his best mathematical work in peace and quiet, ideally while lying in bed in some isolated cabin. And yet he spent many years traveling around Europe in search of adventure, and finding plenty. In 1620, he took part in the Battle of the White Mountain, which was fought near Prague (see right). It was one of the most important battles in the Thirty Years' War, a power struggle between Catholics (including Descartes) and Protestants, and the victorious Catholics redrew the map of Europe as a result of the war. Descartes was a skillful swordsman and in 1621, while the sole passenger on a boat, he needed his swordsmanship when the crew tried to rob and murder him. During his wandering years, Descartes kept in touch with scientific developments in Europe by post, and one of his correspondents was Princess Elizabeth of Bohemia (now part of the Czech Republic). They corresponded about math and philosophy for many years. A later correspondent, Queen Christina of Sweden, was so impressed by Descartes' skills that

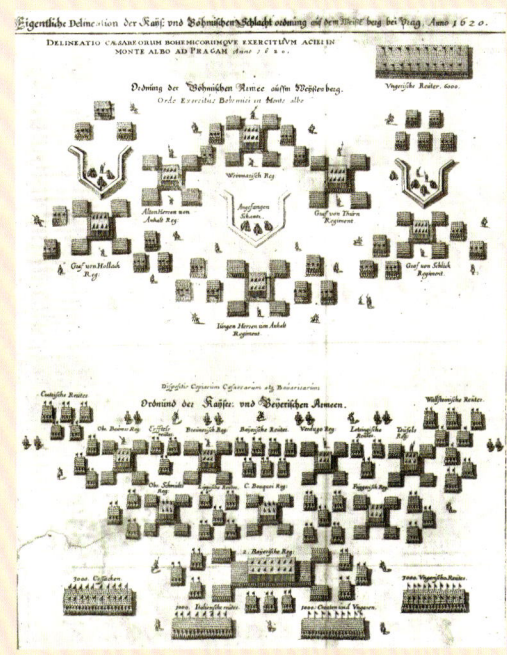

in 1646 she invited him to come to her court in Stockholm so that he could teach her math and philosophy face-to-face (see picture, left). Descartes was very reluctant to go to such a cold country and put her off for three years. But Christina was a very determined woman, and, when she sent a ship with one of her admirals on board to collect him, he felt he could no longer refuse. He arrived in Stockholm in October 1649. Queen Christina was a lot younger and fitter than Descartes (she was 22 when they met, and he was 53), and insisted that they had their daily interviews at 5 o'clock in the morning, in a freezing library. For Descartes, this was even worse than he had anticipated. Nevertheless, he did as she requested for as long as he could. But, after just a few months in Stockholm he caught pneumonia, and died in the bleak February of 1650.

TRANSFORMATIONS

Algebraic geometry can be used to transform shapes. A circle with its center at the coordinates (0,0) is defined by the formula $x^2 + y^2 = r^2$, where r is its radius. The red circle here has a radius of 1, so its equation is $x^2 + y^2 = 1$. Changing the circle's shape is equivalent to modifying the equation; to squash it in the horizontal direction, we multiply the x in the equation by a constant; the equation of the blue shape is $3x^2 + y^2 = 1$. To make the circle larger, we increase the value of r: The green circle's equation is $x^2 + y^2 = 2$. And to move the circle sideways, we just add or subtract a constant to all the x-values: The equation of the orange circle is $(x - 2)^2 + y^2 = 1$.

$x^2 + y^2 = 1$

$3x^2 + y^2 = 1$

$x^2 + y^2 = 2$

$(x-2)^2 + y^2 = 1$

is a map in which the vertical positions are the y-values in an equation and the horizontal ones are x-values; the coordinates are values of x and y. An equation like $y = 2x + 1$, if fed with a series of x-values, will give a table like this:

x	y
−2	−3
−1	−1
0	1
1	3
2	5

Graph 1

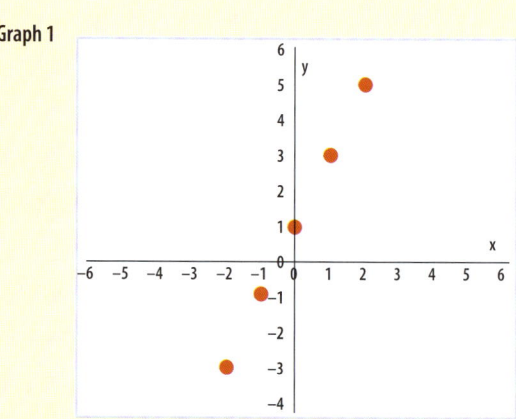

Graph 2

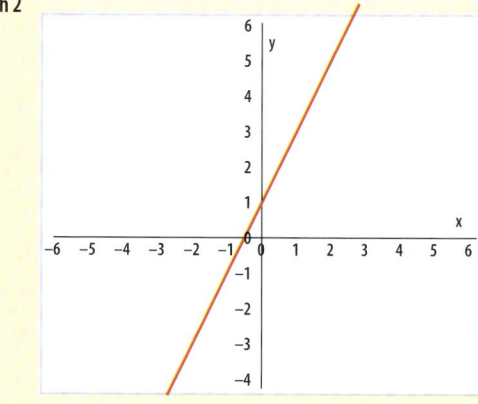

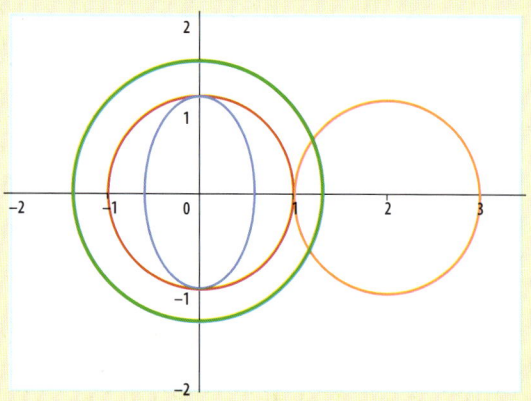

And, by treating these values as coordinates, we can plot them as Graph 1. But there is nothing special about those particular x-values, they are just ones we happened to choose; a better graph would show the y-values corresponding to all possible x-values in the range we are plotting. That would look like Graph 2.

A new card of math

So, now we have plotted an entire mathematical equation, rather than just some solutions to it. This was a major step forward. On the one hand, it provided a new method of studying algebra, by drawing graphs (see box, left). And on the other hand, it meant that geometry could be described by algebra (see box, right). In fact, Descartes showed that any geometrical problem could be converted to an algebraic one.

This system—Cartesian coordinates—is named in honor of Descartes, from the Latinized version of his name, Cartesius. Such a powerful linking-up of two areas of mathematics led to many new developments and discoveries far beyond lines and points on graph paper.

SEE ALSO:
▶ The Algebra of Shapes, page 34
▶ Unreal Numbers, page 74

BROUWER'S FIXED POINT THEOREM

Brouwer's theorem is a complicated one to prove, or to state fully, but very roughly it says that, with some exceptions, two versions of the same thing share a fixed point. It's said that Luitzen Brouwer came up with the idea when he was stirring his coffee, and one of the things his theorem proves is that, however much you stir the coffee, there will be a molecule that ends up exactly where it started. It also means that, if you had two copies of this book, tore this page out, crumpled it up, and dropped it on this same page on the open book, at least one letter on the crumpled version would lie directly over that same letter on the undamaged version. (Though this doesn't work if you spill your coffee on it or rip the page.)

Don't try this at home.

Fermat's Last Theorem

THE WHOLE OF CALCULUS IS BASED ON THE CONCEPT OF INFINITESIMALS. To find the area under a line, we integrate the equation for that line, and this involves dividing the area into many tiny parts and then adding them all up.

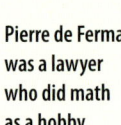

Pierre de Fermat was a lawyer who did math as a hobby.

These tiny parts are infinitesimals. An infinitesimal is something that is too small to have a measurable area and yet is not zero, and a proper definition had to wait until the 19th century (see page 167). Until then, making use of such a mysterious and vague idea in such a precise and clear subject as mathematics discouraged many people. But not Pierre de Fermat. Fermat was born in France sometime near the beginning of the 17th century, and, after training to become a lawyer, he settled down for the rest of his life as a member of the court of appeal in Toulouse. Because his job gave him so much power, he was discouraged from socializing much, in case he was tempted by bribery to give people an unfair advantage in the courts. Being a very law-abiding person, he no doubt did just as he was told, and, as a result, had plenty of time on his hands. This

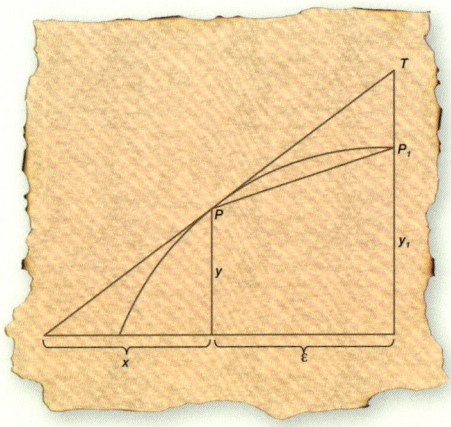

Pierre de Fermat anticipated calculus with his approach to finding the tangent line to a given curve.

may be partly why he took up mathematics; although he published very few papers, he corresponded with many of the mathematicians of his time and thus kept up an active social life through his letters, without venturing on subjects which might affect his job.

Number puzzles

Judging by Fermat's mathematical letters, he also had a great deal of fun in using mathematics to tease his friends. For instance, the sequences of square numbers (1, 4, 9, 16, 25, 36 …) and cube numbers (1, 8, 27, 64, 125 …) have a pair of terms which are just one number apart: $5^2 = 25$ and $3^3 = 27$ are separated by the number 26. This is the only case where a square number and a cube number have just one number between them, although that fact is very hard to prove. Fermat managed it, and challenged other mathematicians to also find a proof. But no one could.

Certain terms

Fermat's trained and brilliant legal mind particularly appreciated the precision of symbolic language, and as a result he was inspired in particular by the works of Viète (page 82) and Diophantus (page 46). One of his most important discoveries was a method of finding the maxima and minima of curves, and this method, and his investigations of the use of infinitesimals, helped Newton and Leibniz to develop calculus (see page 110). Fermat relied on a concept called adequality to "prove" his methods, claiming that he learned about adequality from Diophantus's *Arithmetica*. But in fact Diophantus does not use the term, and even now mathematicians cannot pin down quite what Fermat meant, a fact which Fermat would no doubt find very amusing.

Margin call

By far his most famous discovery almost certainly wasn't a discovery at all. It was so secret, it was only found after his death in 1660, by his son Samuel. Samuel was going through his father's

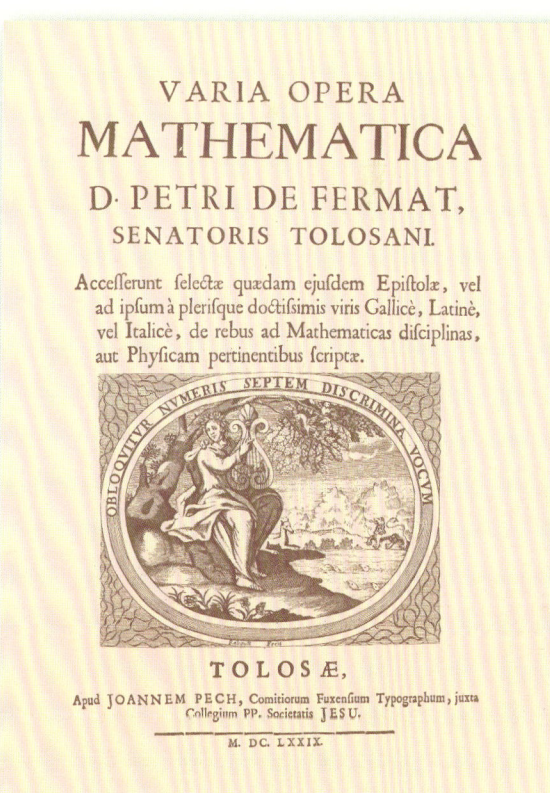

The title pages of Fermat's *Varia Opera Mathematica*, which makes no mention of the writer's most famous idea.

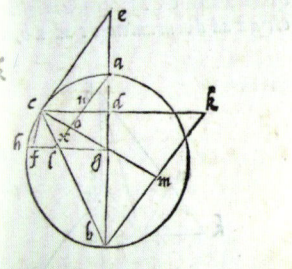

Fermat had a habit of making notes in the margin, as seen here in his copy of Apollonius's *Conics*.

copy of Diophantus's *Arithmetica*, when he found this note in the margin:

> "It is impossible to separate a cube into two cubes, or a fourth power into two fourth powers, or in general, any power higher than the second, into two like powers. I have discovered a truly marvelous proof of this, which this margin is too narrow to contain."

In other words, if you take Pythagoras's theorem, $a^2 = b^2 + c^2$ and replace the 2s with 3s, the resulting formula $a^3 = b^3 + c^3$ is incorrect. There are no natural numbers for which it is true. In fact, Fermat claimed, whatever numbers you replace the 2s with, you will get an incorrect equation. Mathematically, we can write this as $a^n \neq b^n + c^n$ ($n \neq 2$).

Wiles's proof

Many great mathematicians attempted to solve Fermat's Last Theorem, as it became known, and the one who finally managed to do so was Andrew Wiles, who published the proof in 1995, after working on it for 30 years, since the age of 10! Wiles's proof was based on the Taniyama-Shimura conjecture, which states that two very different kinds of mathematical entity are actually closely linked.

On one hand, there are equations of the form $y^2 = x^3 + ax^2 + bx + c$, which are called elliptical equations. Opposite is an example, the curve of the elliptical equation $y^2 = x^3 + 2x^2 + 2x + 2$.

Andrew Wiles proved Fermat's Last Theorem in 1995, after more than 300 years of struggle by mathematicians around the world.

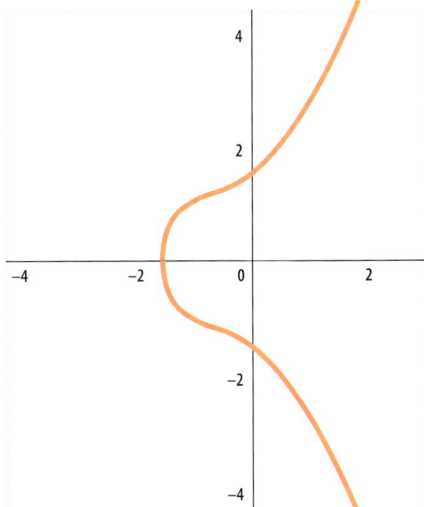

An example of an elliptical curve, a crucial part of Wiles's proof of Fermat's Last Theorem.

On the other hand are modular forms. These can't be explained without very advanced mathematics, but each is a four-dimensional shape that is highly symmetrical. It is the very high degree of symmetry which makes modular forms interesting (see page 142 for more on the deep significance of symmetry).

The Taniyama-Shimura conjecture says that every modular form corresponds to an elliptic curve. Evidence for this is that any modular form can be reduced to a sequence of numbers, and that sequence can be used to plot an elliptic curve.

SEE ALSO:
▸ Cubics, page 64
▸ Algebraic Geometry, page 92

PROOF AT LAST

Wiles's proof is long, complex, and includes many areas of mathematics (some of which he had to develop for himself). In very rough outline, it goes like this:

1. If Fermat's Last Theorem is false then, for some unknown value of n (other than n=2),
$a^n = b^n + c^n$

2. $a^n = b^n + c^n$ ($n \neq 2$) can be transformed into an elliptic curve defined by
$y^2 = x^3 + (a^n - b^n)x^2 + a^n b^n$

3. There is NO modular form corresponding to
$y^2 = x^3 + (a^n - b^n)x^2 + a^n b^n$

4. But, as we know from Taniyama-Shimura, this is impossible.

5. So
$y^2 = x^3 + (a^n - b^n)x^2 + a^n b^n$
is false.

6. But, as we said in stage 2.,
$y^2 = x^3 + (a^n - b^n)x^2 + a^n b^n$
is equivalent to
$a^n = b^n + c^n$ ($n \neq 2$),
so
$a^n = b^n + c^n$ ($n \neq 2$) is therefore false, too.

7. But "$a^n = b^n + c^n$ ($n \neq 2$) **is false**" is Fermat's Last Theorem.

8. So Fermat's Last Theorem is proved true.

Pascal's Triangle

LIKE DESCARTES, BLAISE PASCAL WAS AS INTERESTED IN PHILOSOPHY AND PHYSICS as he was in mathematics. He is remembered today for a powerful pattern of numbers.

Pascal's father was a tax officer, and it was he who taught Blaise about mathematics. A large part of Pascal senior's job was carrying out calculations—many, many calculations. Pascal resolved to help his father by building a machine to perform those calculations for him. He began work on his mechanical calculator when he was 19; it took him untold hours, three years and fifty part-complete machines, but the end result was one of the first mechanical calculators ever built, soon to be called the Pascaline.

Games of chance

Pascal was one of the many people with whom Fermat (see page 98) exchanged letters about mathematics, and together they developed many of the fundamental concepts in probability theory. In those days, the main driving force behind probability was gambling. Rich people (and some not-so-rich ones, too) would bet large sums on all sorts of things, including the way cards, dice, or coins fell, and it was important to know the likelihood of different outcomes. For instance, if I toss three coins, what is the likelihood that I will get two heads (H) and a tail (T)? The simplest way to find out is to list all possible outcomes. There are eight:

In later life, Blaise Pascal turned away from math and science to devote his life to thinking about life—and death.

HHH, HHT, HTH, THH, TTH, THT, HTT, TTT. Then select the outcomes that give me what I want, which are these three:
HHT, HTH, THH.

And divide the numbers: 3/8. So, I have a 3 in 8 chance of getting the outcome I want. This can be expressed as a fraction: 3/8=0.375, or as a percentage: 0.375×100=37.5%. To find out the chance of getting all tails, all heads, or two tails and a head we proceed in the same way, and the probabilities turn out to be:

All heads 1/8.
Two heads and a tail 3/8.
Two tails and a head 3/8.
All tails 1/8.

Below: One of the nine Pascaline calculators that still survive.

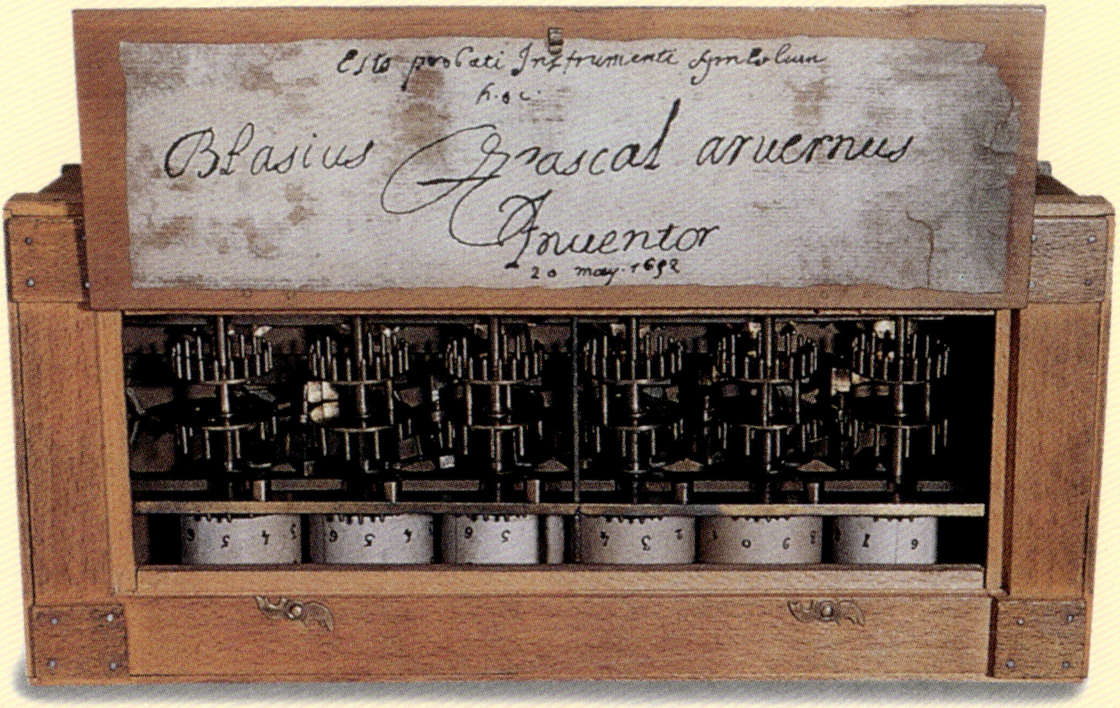

Above: Some of the permutations of three coin tosses.

Building a system

That was easy enough, but it would be quicker and more reliable to work out the possible outcomes of tossing coins from a formula, rather than by listing them all. To do this, Pascal used a technique which had long been known in mathematics (though he probably discovered it for himself). He simply wrote down a triangle of 1s:

1

1   1

And then built up a triangle like this one on the right, in which each number is the sum of the number above it to the right, and above it to the left. (If there is no number shown in one of these places, imagine there is a zero there). And we can continue as long as we wish—we have produced the first eight rows (0 to 7) below.

The sum of each row is on the right and the number of the row is on the left. The row of green numbers here represents the probabilities of the 8 possible outcomes of tossing 3 coins, as we found before, and tossing 4 coins gives the chance of one of 16 possible outcomes, as shown

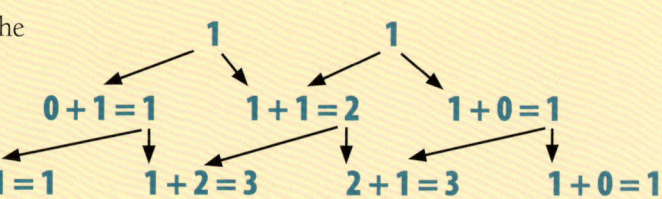

DIAGONAL PATTERNS

1. The second diagonal lists the natural numbers.

2. The third diagonal contains triangular numbers, which are the numbers of spheres needed to form triangles.

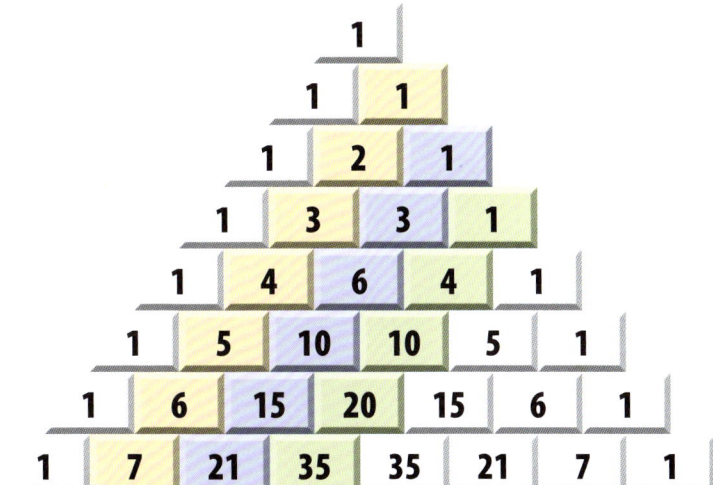

3. And the fourth diagonal is the tetrahedral numbers; the number of spheres needed to build tetrahedra.

POWERS OF 11

The triangle's rows can be read off as powers of 11. Although this seems to break down from the fourth row, actually the results still make sense in terms of the expression on the right.

	Value	Pascal triangle row	Expression
11^0	1	1	$1(10^0)$
11^1	121	1,2,1	$1(10^2) + 1(10^1) + 1(10^0)$
11^2	1,331	1,3,3,1	$1(10^3) + 1(10^2) + 1(10^1) + 1(10^0)$
11^3	14,641	1,4,6,4,1	$1(10^4) + 1(10^3) + 1(10^2) + 1(10^1) + 1(10^0)$
11^4	161,051	1,5,10,10,5,1	$1(10^5) + 1(10^4) + 1(10^3) + 1(10^2) + 1(10^1) + 1(10^0)$
11^5	1,771,561	1,6,15,20,15,6,1	$1(10^6) + 1(10^5) + 1(10^4) + 1(10^3) + 1(10^2) + 1(10^1) + 1(10^0)$

A simple way to think of this is by imagining that all two-digit numbers extend into the boxes on their left, and that the part of the number that has intruded into that box is then added to the number that was there already. So, for line 5, the values look like this:

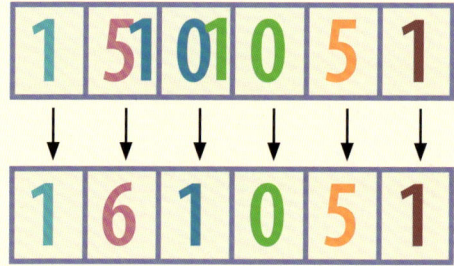

in row 4. Exploring further, there is a 1 in 16 chance of four heads, a 4 in 16 chance of three heads and one tail, a 6 in 16 chance of three heads and three tails, a 4 in 16 chance of one head and three tails, and a 1 in 16 chance of all tails.

Power series

The right-hand column (the sums of the numbers in the rows) is a series. In this case, it shows the powers of 2: $2^0=1$, $2^1=2$, $2^2=4$, $2^3=8$, $2^4=16$, and so on, so it is called a power series. Just as the Pascaline took a lot of the labor out of arithmetic, so Pascal's Triangle, as the system is now known, simplifies tedious work in mathematics, and makes errors less likely, too. For instance, to expand (or multiply out) the expression $(x+y)^2$ is quite quick and easy: $x^2+2xy+y^2$. But what about the expansion of $(x+y)^5$?

We can find out from the triangle, in which line 2 lists the coefficients of our first expansion, $(x+y)^2$, as $1x^2 + 2xy + 1y^2$. So, to find the coefficients of x+y to the fifth power, we just read them off from line 5: $(x+y)^5 = 1x^5 + 5x^4y + 10x^3y^2 + 10x^2y^3 + 5xy^4 + 1y^5$ (In case you are wondering about line zero, that works, too: $(x+y)^0 = xy^0 = 1$.)

Because there is a pair of variables in the brackets, and the prefix "bi" means "two," these expansions are called binomial expansions. Pascal found a great many other patterns in the triangle, and later mathematicians found even more (as shown in the boxes throughout this chapter).

Almighty clash

Though Pascal was a very talented mathematician, by far the most important thing in his life was religion. His parents were members of an extreme religious sect called Jansenism, and Pascal decided that he had been personally chosen by God to make life miserable for Jesuits, who he believed were destined to go to hell. Unfortunately, René Descartes had been brought up by Jesuits, so when the two men—perhaps the greatest mathematicians of their age—finally met, it was not a very successful event.

Near-death experience

Pascal's religious beliefs were strengthened when he narrowly escaped death in 1654. The horses that were pulling his carriage bolted and ran off the edge of a bridge. The "traces" which connected them to the carriage, broke at the last moment and the carriage, and Pascal, survived.

SEQUENCES AND COMBINATIONS

The Fibonacci sequence is in the triangle, too: adding the numbers on the diagonals shown below gives:

1, 1, 1 + 1, 2 + 1, 1 + 3 + 1 …

Which is **1, 1, 2, 3, 5 …**

The triangle can determine combinations: Let's say you can choose any two days off in a week. How many different options do you have? To find out, go to row 7 (because there are 7 days in a week; count the top row as row 0) and go along 2 places (for your 2 days off; count the first place as 0). The number there is 21, so there are 21 choices (Sa+Su, Sa+M, Sa+Tu, Sa+W, Sa+Th, Sa+F, Su+M, Su+Tu, Su+W, Su+Th, Su+F, M+Tu, M+W, M+Th, M+F, Tu+W, Tu+Th, Tu+F, W+Th, W+F, Th+F).

PASCAL'S WAGER

To philosophers, Pascal's name is better remembered for his "wager"(or bet) than for his triangle. Pascal's Wager has various forms, and one of them is "If God exists and I pray to him, then I might go to heaven. If God doesn't exist and I pray to him, there's no harm done. So I might as well pray to him."

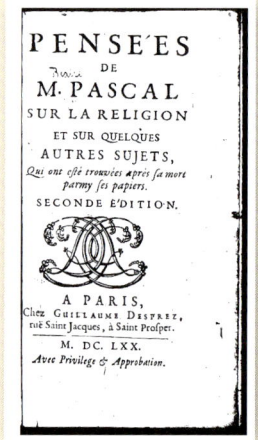

Pensees (Thoughts), a self-explanatory book by Blaise Pascal, contains his wager on prayer.

For ever afterwards, Pascal was haunted by the idea of the fall, even indoors. Sometimes his friends would have to place a chair where he thought the imaginary bridge parapet was, to reassure him that it was a terrible delusion.

Later life

Pascal's beliefs were so extreme that he began to fear that even doing mathematics would displease God, but one night he had a toothache and couldn't sleep. He tried some mathematics to take his mind off the pain, and the toothache faded away. He decided that this showed that mathematics must be an acceptable activity, and went on to make further breakthroughs. Pascal died at just 39, leaving behind some interesting work in physics (especially fluid motion, which today is one of the most active areas of mathematical research, see page 119), and in philosophy, most notably about the mathematical chances of going to hell (see box, left).

SEE ALSO:
▶ Sequences and Series, page 66
▶ The Pythagoreans, page 26

FRACTAL PATTERN

As with so many major new ideas in mathematics, not all of the patterns in the triangle could be understood at the time. Long after Pascal's death, a number of mathematicians began exploring fractals. A fractal is a pattern that looks the same at a wide range of magnifications. One example is the coastline of a country, which is crinkly however you look at it. If you took a photo of a typical piece of coastline looking straight down from heights of 1 mile, 1 yard, 1 foot, and 1 inch, the line where sea meets land would look very similar.

The fractal in this Pascal's Triangle is itself a triangle: Sierpinski's Triangle.

This is a pattern of triangles of all sizes which looks the same at a wide range of scales, and it is generated simply by coloring in the odd numbers in Pascal's Triangle.

This vaguely triangle-like pattern is repeated for larger Pascal's Triangles, as if we are zooming out.

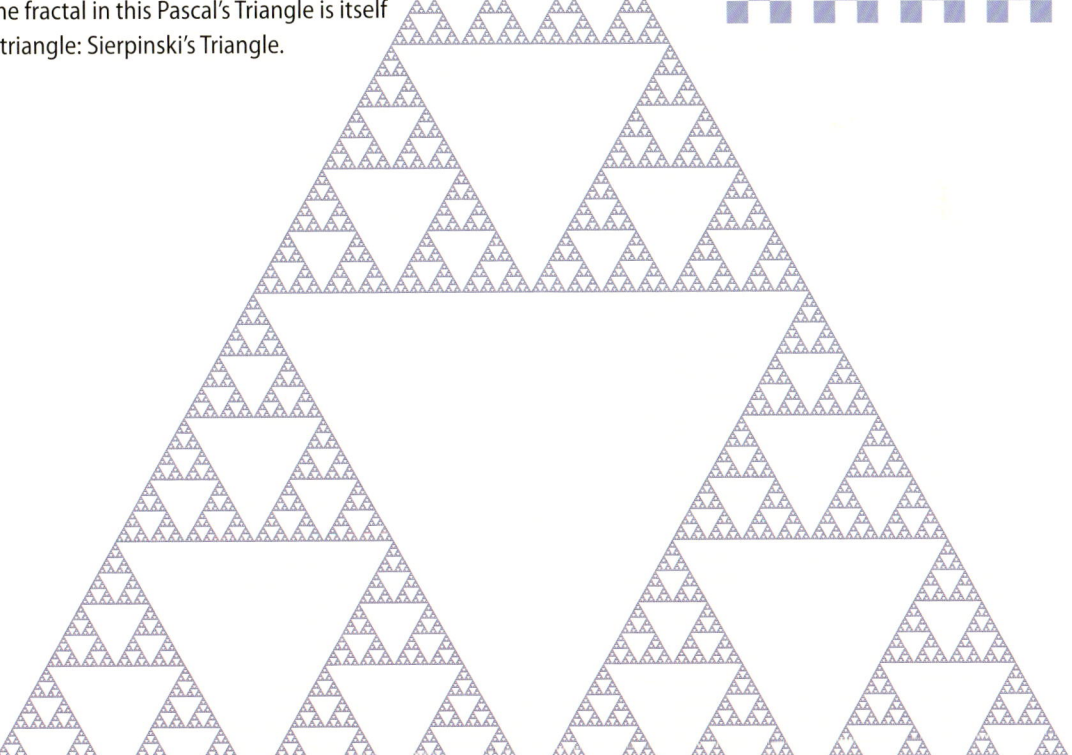

Calculus

CALCULUS IS PROBABLY THE SINGLE MOST IMPORTANT TOOL IN BOTH MATHEMATICS AND SCIENCE. Its job is to deal with things that change, from a herd of wildebeest to the temperature of a chemical reaction.

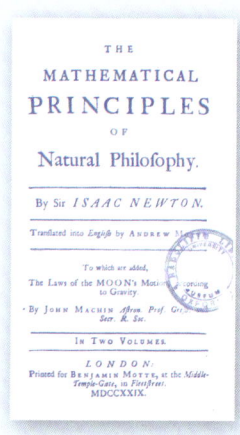

The Principia (Principles) was the result of work done when Isaac Newton was secluded at his country house to avoid the plague that ravaged England.

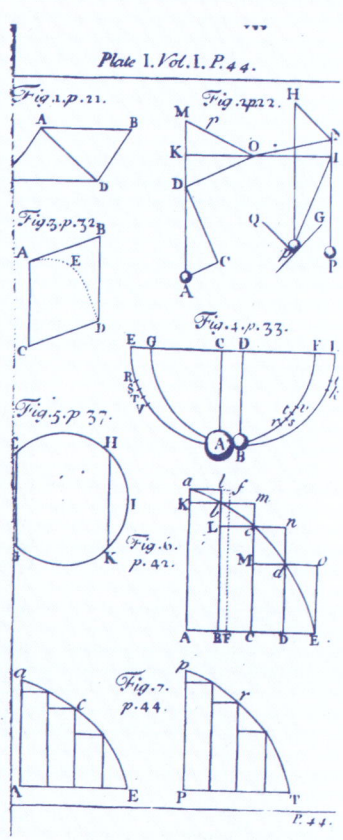

While many mathematicians made discoveries that can now be seen as precursors to calculus, there is no doubt as to who invented the true version. It was Isaac Newton. Or Gottfried Leibniz. The trouble was that Newton was secretive even by the standards of 17th-century mathematicians. He eventually published his masterpiece in 1687. *The Principia*, or *The Mathematical Principles of Natural Philosophy*, is perhaps the most important science book ever written ("natural philosophy" is what we would call physics today). Based on his laws of motion and gravitation, Newton was able to explain and predict precisely the motions of the Moon, planets, and comets, as well as those of falling or thrown objects on Earth (other than the effects of air resistance). In the book, he uses Greek geometrical methods to prove his many powerful results, but most historians believe that in fact he used calculus to work them out first. He probably recalculated them geometrically because he knew that no one would argue with the Greek-style demonstrations, and the physics in the book

Leibniz also developed Pascal's idea of a mechanical calculator, building a machine that could multiply as well as add. Its design was still in use in the 1940s.

was quite challenging enough, without worrying people about a new kind of mathematics, too. Unfortunately for Newton, the person who is credited with a discovery is the one who tells people about it first. And that was definitely German philosopher and mathematician Gottfried Leibniz, who published his version of calculus in 1700.

A complete system

Both men were certainly geniuses, and both had grand plans. Newton wanted to make a complete mathematical theory of the Universe, unlock the secrets of eternal life, and decipher what he thought was a hidden meaning of the Bible. Meanwhile, Leibniz wanted to develop a complete philosophical theory of the world, develop a logical language that would mean that all disputes and arguments could be resolved by calculation, and put an end to all religious wars.

It is now well accepted that both men discovered calculus independently, but at the time—and for long after—arguments raged between British and continental European scientists as to who should be given credit for the discovery.

Differentiation

Differentiation allows rates of change to be determined. For a function of the form $y=ax^n$, the differential is

$$\frac{dy}{dx} = nax^{(n-1)}$$

So, for example, a quadratic equation like $y = 5x^2 - 5x + 12$

Can be differentiated like this

$$\frac{dy}{dx} = 5 \times 2 \times x^{(2-1)} - 5 \times 1 \times x^{(1-1)} + 0$$

$$\frac{dy}{dx} = 10x^1 - 5x^0 + 0$$

Because anything raised to the 1st power just remains unchanged, $10x^1 = 10x$

And, since anything raised to the power of 0 is 1, $5x^0 = 5$.

So we end up with

$$\frac{dy}{dx} = 10x - 5$$

In case you're wondering what has happened to the 12, this can still be regarded as the outcome of applying the $dy/dx = nax^{(n-1)}$ formula. Since $x^0 = 1$, in the original quadratic we can regard the 12 as $12x^0$. Feeding that into our formula gives $10 \times 0 \times x^{(0-1)}$, which equals 0.

The formula $dy/dx = nax^{(n-1)}$ is based on the idea that a curve can be thought of as a multitude of short straight lines. In the same way that a straight line has a single slope (see page 92), a curved line has many, in the sense that at any point on the curve a straight line can be drawn which just skims the line without passing beneath it. This is called the tangent, and its slope is the slope of the curve at that point. We can check this by differentiating the equation of a curve, say $y = x^2$. The formula gives $dy/dx = 2x$. So, at the point defined by $x = 1$, the slope is $2 \times 1 = 2$, and this makes sense in terms of a graph of the $y = x^2$ curve; the slope of a tangent to the curve at the point $x = 1$ is indeed 2.

We know that a straight line can be described by this equation: $y = mx + c$ and the slope of that line is

$$m = \frac{(y_2 - y_1)}{(x_2 - x_1)}$$

Now, because the curve slope is different at every point, we can't choose just any values of x_1 and x_2. If they are too far apart, the line that passes through them will cut across the curve. We want x_1 and x_2 to be so close that the very short part of the curve that joins them is as similar to a straight line as makes no difference. Let's define this very short distance by using the letter d. So, a tiny change in x will be dx and a tiny change in y will be dy. We can now define our slope equation as

$$m = \frac{(y_1 - (y_1 + dy))}{(x_1 - (x_1 + dx))}$$

Let's take the equation $y = x^2$. To find the expression for its slope(s), we replace the ys in our equation for m with x^2s

$$m = \frac{(x_1^2 - (x_1 + dx)^2)}{(x_1 - (x_1 + dx))}$$

Multiplying this out gives

$$m = \frac{(x_1^2 - (x_1^2 + x_1\,dx + x_1\,dx + dx^2))}{-dx}$$

Which is

$$m = \frac{-2x_1\,dx + dx^2}{-dx}$$

Now, we have said that dx is very tiny indeed. So, squaring it will make it even tinier (just as the square of one millionth is one trillionth). So, we can say that $m \approx 2x_1$ ($\approx$ means "approximately"). We would have got this expression whatever value of x we had chosen,

The Royal Society of London debated who invented calculus; the chairman of the discussions was Isaac Newton, who decided that the true inventor was definitely himself.

so rather than writing x_1, we can just write x, to give $m \approx 2x$. This is the differential we want so we can conclude that, for $y=x^2$

$$\frac{dy}{dx} \approx 2x$$

By carrying out similar analyses on curves of x^3 and other powers, we get the general formula

$$\frac{dy}{dx} \approx nax^{(n-1)}$$

We just need to deal with that ≈ sign, and the next part of this argument is one which led to decades of debate (which we'll take up on page 166). What we say is this: "dx is a very small change in x, so small that, in fact, we can make dx as close to zero as it can be, without actually becoming zero." We can call this "tending to zero" or "becoming zero in the limit." The reason dx can't quite be zero is that the fraction dy/dx would then be ⁰/₀, which is meaningless. Because dx is practically zero, $(dx)^2$ is so tiny that we will be safe to say that it IS zero. So, we will say $(dx)^2 = 0$. In which case we can get rid of that ≈ sign:

$$\frac{dy}{dx} = nax^{(n-1)}$$

This argument may not strike you as particularly convincing, and many mathematicians felt just the same. On the other hand, the differentiation formula certainly works well. We can see this by checking it against something we already know (don't forget: differentials don't only mean slopes to lines). For example, the differential of the formula for the area (A) of a circle gives the

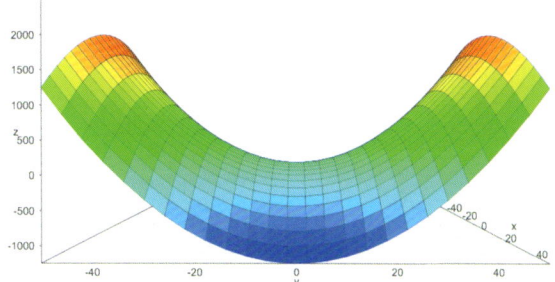
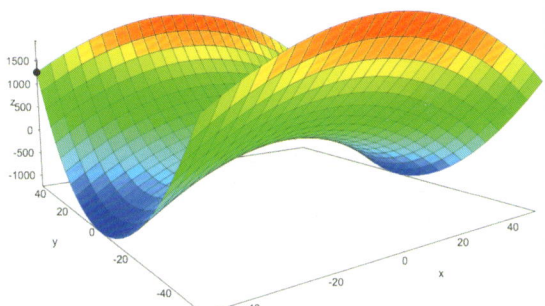
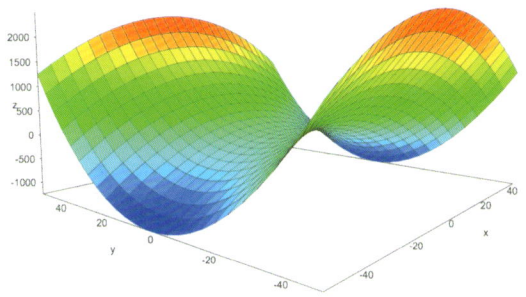

For differentiating a three-dimensional surface, we add a third variable, z (see page 176).

correct formula for the circle's circumference (C), as we can see:

$$A = \pi r^2$$

$$C = \frac{dA}{dr} = 2\pi r^{2-1} = 2\pi r^1 = 2\pi r$$

So, we can be reassured that differentiation works, even if the basis of it may be questionable.

Integration

Integration is the reverse of differentiation. Among other things, it is used to find areas under curves. For an equation of the form $y = ax^n$ the integral is

$$\int ax^n dx = \frac{ax^{(n+1)}}{(n+1)} + c$$

There are a couple of points to make here. The formula does not apply if $n = -1$, and that c at the end is an unknown constant. (Since the differential of a constant is zero, so the integral of zero is a constant). This constant that appears when we integrate a function is very inconvenient, but fortunately we can get rid of it when we actually want to do something useful. Let's say we have a spacecraft, the velocity of which is increasing at a rate of 0.4 feet per second every second. That is, $v = 0.4t$

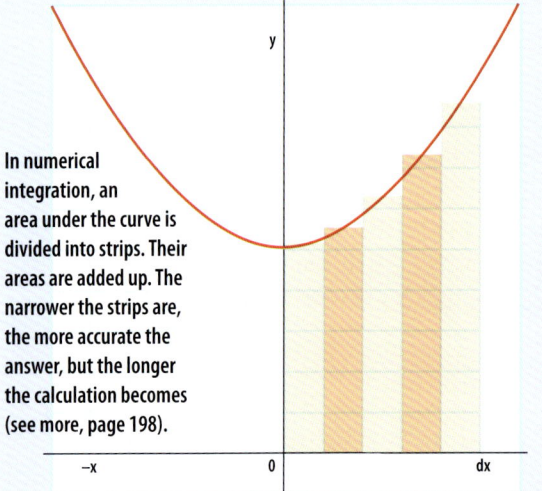

In numerical integration, an area under the curve is divided into strips. Their areas are added up. The narrower the strips are, the more accurate the answer, but the longer the calculation becomes (see more, page 198).

Say we want to answer the question "How far has the spacecraft traveled?" Well, acceleration is the rate of change of velocity, and velocity is the rate of change of position (sometimes referred to as displacement, or simply distance). Differentials are just rates of change, so acceleration is the differential of velocity and velocity is the differential of position. Also, since integrals are the opposite of differentials, so position is the integral of velocity, and velocity is the integral of acceleration.

So, to find how far something has traveled, we find the integral of the equation for its velocity. We can plug the velocity equation into the standard formula, integrating with respect to t (for time), rather than x.

$$\int at^n \, dt = \frac{at^{(n+1)}}{(n+1)} + c$$

The velocity equation is $v = 0.4t$. To be clear about what happens next, we can write this with the variable in the form of a power, i.e. t^1 (of course, t^1 is the same as t, so the 1 isn't usually written, but we'll need to be aware of its presence on the next line). So the coefficient, represented by a, is 0.4. In which case we have

$$\int 0.4t^1 \, dt = \frac{0.4t^{(1+1)}}{(1+1)} + c$$
$$= \frac{0.4t^2}{2} + c$$
$$= 0.2t^2 + c$$

Which is not a lot of use, because the c could be anything.

Without Newton and Leibniz, controlled space flight would be impossible.

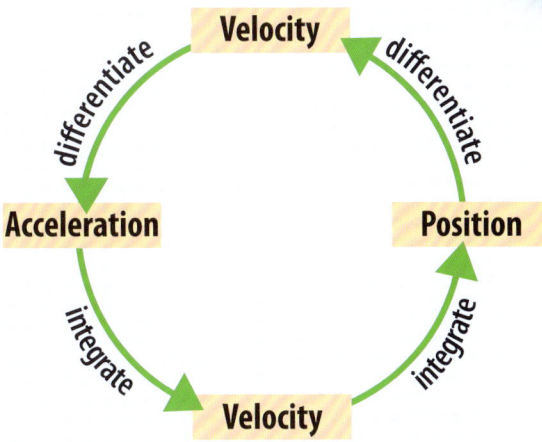

But when we look back at our question, it becomes clear why we get this unhelpful answer. It is because the question is ambiguous. There's no point asking "How far has that rocket traveled?" unless we know how long it has been traveling for. Once we've found out how long ago it was launched, we can ask a question that we can reasonably expect to answer. So, we might ask "How far has that rocket traveled in the 10 minutes since it launched?"

We are now posing a question with a definite answer instead of an indefinite one, and similarly, we can calculate what is called a definite integral, rather than an indefinite one. The definite integral in this case is

$$\int_{t=0}^{t=600} 0.4t\, dt$$

The 600 is just those 10 minutes converted into seconds. The 0 is just reminding us that the rocket launched at a time we are calling zero, for simplicity. (The small ts are called limits). The integration proceeds as before, but the answer ends up with square brackets around it $[0.2t^2 + c]_0^{600}$ We evaluate it by replacing the variable (t) with each limit (600 and 0) in turn, and then finding the difference
$= (0.2 \times 600^2 + c) - (0.2 \times 0^2 + c)$
$= 72,000 + c - c$
$= 72,000$

Which is the distance traveled by the spacecraft in feet, since our original velocity was in feet per second. And we've got rid of the inconvenient c without ever having to find out what it was.

See page 176 for advanced integration and differentiation methods in Calculus in Depth.

SEE ALSO:
▶ The Fundamental Theorem of Calculus, page 136

Differential Equations

NATURAL PHENOMENA ARE ALL ABOUT CHANGES, and differential equations are mathematical tools for telling us what changes we can expect in different situations.

Most of science is about studying the way things change, whether those things are stars, moving objects, reacting chemicals, living things, the human mind, or the entire universe. Scientists attempt to define and explain the laws that describe those changes. These laws can then be used to predict how the things they describe will behave in different situations. The most powerful and useful scientific laws are those that can be expressed in mathematical forms, and in many cases those forms are differential equations.

Difference makers

In the simplest kind of differential equation, a differential (that is, something which changes) is set

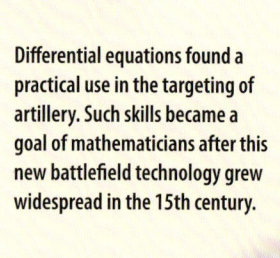

Differential equations found a practical use in the targeting of artillery. Such skills became a goal of mathematicians after this new battlefield technology grew widespread in the 15th century.

equal to a single variable. For instance, a rocket's velocity will change when its jets are thrusting it, and the harder the thrust (which is the variable, T), the more rapidly the velocity will change. Here, the differential is the change in velocity, which we can write as dv/dt). For a particular rocket, perhaps 4 units of thrust (4T) applied for a period of 1 second will increase the velocity by 1 foot per second. We can write this as

$$\frac{dv}{dt} = 4T$$

This is a differential equation. What might we use it for? Well, we can extract another equation from it, one that tells us the velocity of the rocket rather than the change of velocity. To move from velocity to change of velocity, we differentiate, as above. So, to go the other way, from change of velocity to velocity, we integrate.

$$v = \int \text{change of velocity} = \int \frac{dv}{dt} dt = 4Tt + c$$

The problem is the constant of integration, c. But as usual, we can solve problems by asking a more specific question: what is the velocity after applying the thrust for 30 seconds? This is answered by finding the definite integral, like this

$$\int_{t=0}^{t=30} \frac{dv}{dt} dt = [4Tt + c]_0^{30} = 120T$$

Going further

The simple example above only describes a particular kind of rocket, but a few laws of physics are almost as simple, such as the law of gravity. An object close to the Earth and falling toward it will be accelerated by the Earth's gravity (acceleration is the same thing as change of velocity). The gravitational acceleration of Earth (g) is around 32.2 feet per second per second (that means that, if an object is dropped, it will be moving at 32.2 feet per second after the first second, 64.4 after the 2nd second, 96.6 after the third second, and so on). So we can write $dv/dt = g$.

From this differential equation we can derive another equation, that tells us the velocity of the

Isaac Newton used the path of a comet in 1680 (the so-called Newton's comet) to test his new form of mathematics.

falling object, rather than its acceleration. Again, to move from acceleration to velocity, we integrate:

$$v = \int acceleration = \int \frac{dv}{dt} dt = 4Tt + c$$

We next get rid of the constant of integration by asking a more specific question: what is the velocity after falling for 30 seconds? The answer is:

$$\int_{t=0}^{t=30} \frac{dv}{dt} dt = [g+c]_0^{30} = 30g = 294.3 \text{ ft/s}$$

Ordinary or partial

Differential equations are a natural development from calculus, so it's no surprise to find them in the work of both Leibniz and Newton. The first person to actually write one down was Leibniz, and this is it:

$$\int x dx = \frac{1}{2} x^2$$

This is very easy to solve, we just need to differentiate:

$$x = \frac{d(\int x dx)}{dx} = \frac{x^2}{2} + c$$

If only they were all like that, the history of mathematics and physics would have been a lot easier. However, Newton needed differential equations as tools to take him from the laws of motion and gravitation he had discovered to the solution of actual problems (like the speed of the Moon in its orbit). It was he who tried to work out how to solve more complex ones, and it was he, too, who classified them into the two

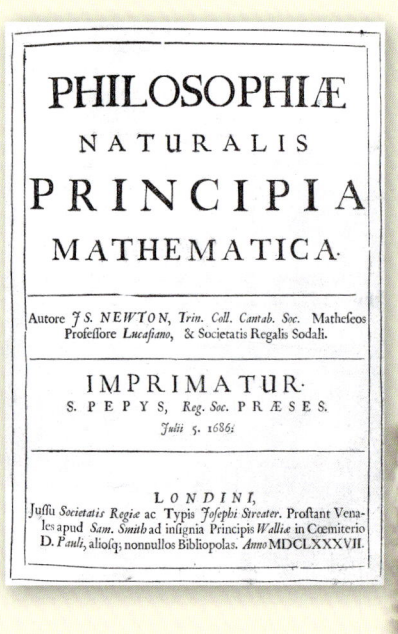

Although Newton used differential equations to develop his new physics, there are none in his in his 1687 masterpiece *The Principia*.

main types we still consider today: Ordinary Differential Equations (ODEs) and Partial Differential Equations (PDEs). The one in the example above is an ODE, and these are often used to study physical phenomena. Nature is, however, pretty complicated, and physical laws usually involve several variables. For instance, any law which relates to positions in space (perhaps one that describes the temperature of the sea, T) will probably depend on the locations of objects in three dimensions (x, y, and z). So, to describe how the temperature varies according to position, we will need three differentials, $\partial T/\partial x$, $\partial T/\partial y$, and $\partial T/\partial z$. And, if we want to study how the temperature changes with time (t), too, we'll need a fourth differential, $\partial T/\partial t$. These are all partial differentials. In most cases, these PDEs are too difficult to solve exactly; in many cases they cannot be solved at all, and in a few cases, we don't even know whether or not they have any solutions.

Moving liquids

One particular PDE has produced a million-dollar question. The behavior of any fluid, however it moves, is determined by the interaction of four forces, related to inertia, pressure, viscosity (thickness), and gravity. They add up like this: inertial forces (which depend on the density of the liquid (ρ) and the velocities in each direction (v_x, v_y, v_z) = − pressures (P) + forces due to

The Anglo-Irish mathematician Sir George Gabriel Stokes. He never actually met the French engineer and physicist Claude-Louis Navier, despite their common interest in the mathematics of fluids.

viscosity (μ) + gravitational force (g)). Conditions in a moving fluid will be different depending on position and time, so the differential equation that describes fluid behavior will be a partial one. In fact, it's most convenient to define three partial differential equations, one for each dimension. These are called the Navier–Stokes equations. They can model almost

$$\rho\left(\frac{\partial v_x}{\partial t} + v_x\frac{\partial v_x}{\partial x}v_y + \frac{\partial v_x}{\partial y} + v_z\frac{\partial v_x}{\partial z}\right) = -\frac{\partial P}{\partial x} + \mu\left(\frac{\partial^2 v_x}{\partial t^2} + \frac{\partial^2 v_x}{\partial y^2} + \frac{\partial^2 v_x}{\partial z^2}\right) + \rho g_x$$

$$\rho\left(\frac{\partial v_y}{\partial t} + v_x\frac{\partial v_y}{\partial x}v_y + \frac{\partial v_y}{\partial y} + v_z\frac{\partial v_y}{\partial z}\right) = -\frac{\partial P}{\partial y} + \mu\left(\frac{\partial^2 v_y}{\partial t^2} + \frac{\partial^2 v_y}{\partial y^2} + \frac{\partial^2 v_y}{\partial z^2}\right) + \rho g_y$$

$$\rho\left(\frac{\partial v_z}{\partial t} + v_x\frac{\partial v_z}{\partial x}v_y + \frac{\partial v_z}{\partial y} + v_z\frac{\partial v_z}{\partial z}\right) = -\frac{\partial P}{\partial z} + \mu\left(\frac{\partial^2 v_z}{\partial t^2} + \frac{\partial^2 v_z}{\partial y^2} + \frac{\partial^2 v_z}{\partial z^2}\right) + \rho g_z$$

anything that happens to a fluid. Let's say you push a spoon down into a bowl of syrup. The spoon will apply a pressure to the syrup, which will be pushed away from the spoon in all directions. But the syrup, being viscous (thick), will resist that pressure. Also, like any material or object, the syrup will resist being moved due to its inertia. Finally, the spoon will displace some of the syrup, making it rise a little in the bowl. But this will be resisted by the force of gravity, which will try to pull the syrup back down again.

Right: Coffeehouses began to appear in Europe in the 17th century. The flow of coffee inside helped generate many new math ideas—perhaps even the Navier–Stokes equations.

Below: Luckily water is not very dense or viscous, or Navier–Stokes math would make fountains impossible.

Real world applications

The Navier–Stokes equations are used by scientists and engineers in many fields, including weather forecasting, oceanology, vehicle design, seismology, pipeline engineering, wind farm design, and pollution research.

If you tip a spoonful of milk into a cup of coffee, the Navier–Stokes equations allow you to describe the way in which the milk and coffee mix. Or, if you blow out a candle, the equations will model the pattern of smoke that rises from the wick. In addition, if you wanted to design a ship's propeller that would minimize disturbance to the water as it turns, Navier–Stokes should hold the answer.

Real world difficulty

Usually, if you have a differential equation and a question you want to answer, you would extract from that equation whatever formula best answers your question. So, if you wanted to find out how long it would take for a drop of ink to spread throughout the water in a glass, you would solve the Navier–Stokes equations to give you a formula that relates the time of spreading to the size of the glass and the temperature of the water. Then you would feed the actual values of glass size and water temperature into that formula, and read off the answer you want.

Unfortunately, there's a problem. Mathematicians have been unable to find general solutions for the Navier–Stokes equations, despite more than a century of effort. All we have are some equations for particular cases, and some more general, but

The Clay Mathematics Institute has offered hard cash as an incentive to solve the toughest math problems, including the Navier–Stokes equations.

very approximate, formulas. Even these are very useful, but a full solution would be enormously more powerful. However, it's not known whether the equations even have a solution. Instead, they might, under some situations, predict physically impossible situations—probably in the form of infinitely powerful explosions, which fortunately are not a real thing.

A general solution to the equations would be so useful that anyone who finds one will win a million dollar prize: The Navier–Stokes mystery is one of seven "Millennium" problems in mathematics, selected by the Clay Mathematics Institute in 2000. The prize will also be awarded to anyone who proves that the equations are unsolvable. See page 174 for more details.

> **SEE ALSO:**
> ▸ Equations, page 46
> ▸ Cubics, page 64

e

2.71828182845904...

e IS PROBABLY THE MOST IMPORTANT CONSTANT IN MATHEMATICS, and it occurs in many everyday situations, such as your bank balance.

Leonhard Euler (pronounced "oiler") may be the greatest of all mathematicians, and he was certainly one of the most prolific: He wrote tens of thousands of pages of mathematical investigations, making contributions to a wide range of areas, steadily producing work from the age of 19, and continuing despite going blind.

In pursuit of math

In the 18th century, the Paris Academy of Sciences frequently offered prizes for solving scientific and mathematical problems, and in 1727 Euler entered a competition to calculate the best arrangement for masts on a ship. He came second, perhaps partly because he'd never actually seen one.

By this time Euler had been invited to St. Petersburg to take up a post in medicine at the city's Academy, which he accepted despite the fact he knew absolutely nothing about the subject. He learned as much as could before setting off to Russia, but when he got there, political turmoil in the country meant there was no money to pay his salary. The only job he could find was as a ship's doctor, which he landed thanks to his (slight) knowledge of medicine and his (not very relevant) study of ship's masts. Luckily, funding was found for him from the city's academy of science before he actually had to operate on anyone. Better still, the money was for a mathematics post, after all.

Among Leonhard Euler's many books was his *A Method For Finding Curved Lines* of 1744. He was so prolific that not all of his work has been published even now.

From Russia to Prussia

For most of his 4 years at the St. Petersburg Academy, political troubles continued, and Euler had to keep his head down to avoid becoming involved. So he was very pleased to accept an invitation from Frederick II, King of Prussia, to join yet another academy, this time the new Academy of Sciences in Berlin. Not long after he arrived, the Queen Mother, having tried and failed to get Euler to chat with her, asked him why he was so quiet. His reply was, "Because, Madam, I have come from a country where people are hanged if they talk." During the 20 years that he spent at the Berlin Academy, Euler kept producing more and more important mathematics, but he also fell out with the king. The last straw came when the presidency of the academy became vacant. By then, Euler was not only by far the greatest mathematician in the academy, he also carried out a lot of the day-to-day running of the establishment. But the king, having failed to obtain any of his favorite mathematicians for the post, awarded the presidency to himself instead.

To the end

To Frederick's fury, Euler quit, moving back to St. Petersburg, where he had been invited by Empress Catherine the Great to become president of her academy. He continued to work there until he died. Despite losing his sight, he kept up with the latest scientific developments. On the day he died, he calculated the mathematics of the newly invented hot air balloons, and studied the orbit of the recently discovered planet Uranus.

This 1753 map of the world was created by Leonhard Euler while he was working in Berlin, Prussia (now part of Germany). He placed Prussia at the center of the world.

Taking interest

One of Euler's many important discoveries began with a discussion about compound interest with his fellow mathematician and friend, Jacob Bernoulli. If you put $100 in a bank account offering 12 percent interest, how much would you have after one year? The answer is at least $112. That is the amount you would get if you receive the interest only once per year, on the anniversary of your paying-in date.

But you would do better if the interest was paid monthly, because interest in the second month would be paid not only on the $100, but also on the interest that was earned in the first month (so the interest is "compounded" over the months). Over the full year, your money would grow even more, so you're 68¢ better off (see box, right). If the interest is paid weekly, the total will be a little higher. Rather than drawing up a 52-row table to work out how much higher, we can use this expression:

$$p\left(1 + \left(\frac{1}{n}\left(\frac{r}{100}\right)\right)\right)^{nt}$$

Here, p is the original amount invested (p is for "principal"), r is the annual interest rate (expressed as a percentage), n is the number of times per year the interest is paid, and t is the time (in years) over which we want to accumulate the interest. For the simplest case, where there is just a single interest payment at the end of the year, the formula gives:

$$\left(100\left(1 + \frac{1}{1}\left(\frac{12}{100}\right)\right)\right)^{1} = \$112$$

If the interest is added monthly, the total is

$$\left(100\left(1 + \frac{1}{12}\left(\frac{12}{100}\right)\right)\right)^{12} = \$112.68$$

And for weekly interest we would have

$$\left(100\left(1 + \frac{1}{52}\left(\frac{12}{100}\right)\right)\right)^{52} = \$112.73$$

Presumably, the highest total would result if the interest was paid continuously (and there are some bank accounts which do offer such an arrangement).

Leonhard Euler did not find King Frederick the Great of Prussia to be particularly great.

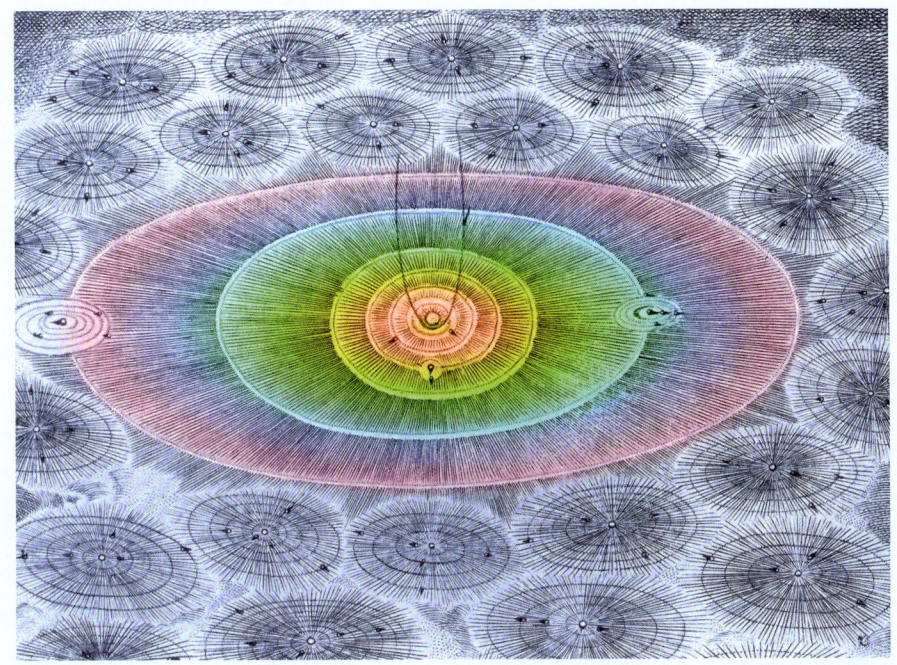

This color-enhanced illustration shows a plurality of worlds as described by Leonhard Euler in 1744. Our Solar System, he said, is just one of many.

We can find out that total by referring to the expression. The only variable that is changing in these examples is n. So, we want to know what happens to

$$\left(1 + \frac{1}{n}\right)^n$$

as n increases. It was Leonard Euler who found the answer to this: as n increases,

$$\left(1 + \frac{1}{n}\right)^n$$

gets ever closer to (or "tends toward") 2.71828 … This number is now known as Euler's number, or just e.

Month	Original investment	Total interest paid to date	Total amount	Interest to be added this month	Total after interest added
1	$100	0	$100.00	$1.00	$101.00
2	$100	$1.00	$101.00	$1.01	$102.01
3	$100	$1.01	$102.01	$1.02	$103.03
4	$100	$1.02	$103.03	$1.03	$104.06
5	$100	$1.03	$104.06	$1.04	$105.10
6	$100	$1.04	$105.10	$1.05	$106.15
7	$100	$1.05	$106.15	$1.06	$107.21
8	$100	$1.06	$107.21	$1.07	$108.29
9	$100	$1.07	$108.29	$1.08	$109.37
10	$100	$1.08	$109.37	$1.09	$110.46
11	$100	$1.09	$110.46	$1.10	$111.57
12	$100	$1.10	$111.57	$1.11	$112.68

So, the highest total is

$$100(e)\frac{12}{100} = \$112.75$$

One very useful characteristic of e^x is that it is its own differential: $de^x/dx = e^x$, and (with the addition of the usual constant of integration) it is its own integral, too: $\int e^x dx = e^x + c$.

The math of growth

e is probably the most important constant in mathematics, and it occurs in many natural situations, too, often in the form of the exponential function, $f(x) = e^x$. This function comes into play when the rate at which something grows or shrinks over time (dA/dt) depends on the amount present ($A(t)$). That is, $dA/dt \propto A(t)$.

An example of this kind of growth is that of bacteria. So long as they have enough food and space, and the right conditions for growth, each individual bacterium will split in two. Its two children will themselves split into two after a time that depends on the type of bacteria, but can range from a few minutes to a few hours. Let's assume four hours. The four grandchildren of the original bacterium will themselves split after

The Swiss mathematician Jacob Bernoulli introduced Euler to the phenomenon of the number e.

Money grows thanks to e (and the financial system).

How it works

Why does
$$\frac{de^x}{dx} = e^x \; ?$$

The value of e^x can be calculated by means of a series:

$$e^x = \frac{x^0}{1} + \frac{x^1}{1} + \frac{x^2}{2} + \frac{x^3}{6} + \frac{x^4}{24} + \frac{x^5}{120} + \ldots$$

The series of denominators here: 1, 1, 2, 6, 24 ... are known as factorials. The factorial of a number n, written $n!$, is given by $n! = 1 \times 2 \times 3 \times \ldots \times (n-2) \times (n-1) \times n$, so $6! = 1 \times 2 \times 3 \times 4 \times 5 \times 6 = 720$. So, the series for e^x is

$$e^x = \frac{x^0}{0!} + \frac{x^1}{1!} + \frac{x^2}{2!} + \frac{x^3}{3!} + \frac{x^4}{4!} + \frac{x^5}{5!} + \ldots$$

If we differentiate e^x, then that is like differentiating the whole series:

$$\frac{de^x}{dx} = \frac{d^{x^0}\!/_{0!}}{dx} + \frac{d^{x^1}\!/_{1!}}{dx} + \frac{d^{x^2}\!/_{2!}}{dx} + \frac{d^{x^3}\!/_{3!}}{dx} + \frac{d^{x^4}\!/_{4!}}{dx} + \frac{d^{x^5}\!/_{5!}}{dx}$$

And those differentials are each given by the usual formula:

$$\frac{dax^n}{dx} = nax^{n-1}$$

If we start working out that series of the differentials, we get:

$$\frac{d^{x^0}\!/_{0!}}{dx} = \frac{0}{1} = 0; \quad \frac{d^{x^1}\!/_{1!}}{dx} = \frac{x^0}{1} = 1; \quad \frac{d^{x^2}\!/_{2!}}{dx} = \frac{2x}{2} = x; \quad \frac{d^{x^3}\!/_{3!}}{dx} = \frac{3x^2}{6} = \frac{x^2}{2}; \quad \frac{d^{x^4}\!/_{4!}}{dx} = \frac{4x^3}{24} = \frac{x^3}{6}; \ldots$$

So, $\dfrac{de^x}{dx} = 1 + x + \dfrac{x^2}{2} + \dfrac{x^3}{6} + \ldots$

Which can be written as

$$\frac{de^x}{dx} = \frac{x^0}{0!} + \frac{x^1}{1!} + \frac{x^2}{2!} + \frac{x^3}{3!} + \ldots$$

Which is just the expression for e^x again.

Graphically, this means that if we calculate the slope of the line $y = e^x$ at any selected point, the slope will have the same value as the position of the point itself:

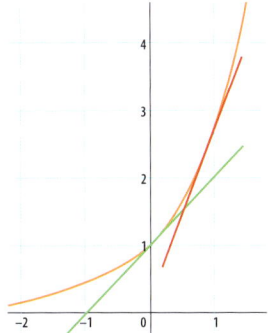

In the graph above, the line is defined by $y = e^x$, and the green line is the tangent to the line at (0,1). It has a slope of 1. The red line is the tangent to the line at the point (1, 2.718), and it has a slope of 2.718.

We can also now explain what the integral of ax^n is when $n = -1$, and why. The graph of $\log_e$ looks like this one on the right:

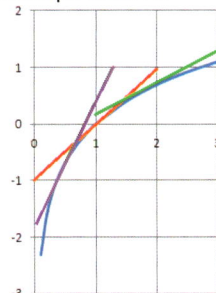

The tangents here are drawn at $x = 0.5$, 1, and 2, and their slopes are 2, 1, and 0.5. So, the differential of $\log_e x$ is $1/x$ (which can also be written x^{-1}).

Conversely, the integral of $1/x$ (or x^{-1}) is $\log_e x + c$, or, more generally, $\int ax^{-1} \, dx = a \times \log_e x + c$

The number e is found in the way bacteria divide.

another four hours, and so on. So, the number of bacteria being produced at any given time depends on the number already present.

So, if the $A(t)$ in our equation represents the number of bacteria present at time t, we can move from our expression $dA/dt \propto A(t)$ to a differential equation by introducing a constant (called a constant of proportionality), say k. So, $dA/dt = kA(t)$.

If we begin with a single bacterium, then how many bacteria will there be after a day? To find out, we integrate the equation, and to do that we can use a method called separation of variables. Variables of one kind are moved to the left side of the equation, variables of the other kind to the right side, and the two sides are then integrated separately:

$$\frac{dA}{A(t)} = kdt$$

$$\int \frac{dA}{A(t)} = \int kdt$$

On the left, we have a $1/A$ term, and whenever we have this form, the integral is always a natural logarithm (see box, right): $\int 1/A(t)\, dA = \ln(A(t))+c$. The integral of the right-hand side is $\int k\,dt = A(t)t + d$. So we have $\ln(A(t)) + c = A(t)t + d$

We can just add the constants together to give another, let's call it "f" (this step is just to simplify the expression): $\ln(A(t)) = A(t)t + f$

Now, we can take the antilogarithm of both sides: $A(t) = e^{A(t)t+f}$. And, since adding powers is the same as multiplying numbers, $A(t) = e^{A(t)t+f} = e^{A(t)t+f} \times e^f$

Again, just to simplify, we will call e^f "g" $A(t) = e^{A(t)t+f} = e^{A(t)t+f} \times e^f = ge^{A(t)t}$

We can simplify this even further, by setting $h^t = e^{A(t)t}$, so $ge^{A(t)t}$ becomes gh^t

And this is what we need, the simplest form of solution of the differential equation for doubling: $A(t) = gh^t$

Into the unknown

We now need to work out the unknown constants, g and h. In our example, the time for the number of bacteria to double is 4 hours. In other words, if $A(t)$ at time zero is, say, 1 (that is $A(0)=1$, then $A(t)$ at time $t = 4$ would be 2 (written $A(4)=2$). So, $A(4)=2A(0)$. Using our equation $A(t) = gh^t$, this means $gh^2 = 2gh^0$. So $h^2 = 2h^0$. But $h^0 =1$, so $h^2 = 4$. Finally, we take the square root of both sides, to end up with $h = 2$. So, we have found one of the unknowns and can say $A(t) = g2^t$

Now, to find g. We said that there was one bacterium at the beginning of the week (which is when $t = 0$), that is $A(0) = 1$. So $A(0) = 1 = g2^0$. So, that means
$$g = \frac{1}{2^0} = \frac{1}{1} = 1$$

So the final form of our equation for these particular bacteria is $A(t) = 2^t$ and the number of bacteria after a day (24 hours) is $A(24) = 2^{24} = 16{,}777{,}216$. This large number just goes to show how rapid exponential growth can be. It also shows that we need to be careful about how well mathematical formulas fit the real world: 16 million or so bacteria are still very little (even a trillion bacteria only weigh about a gram), so this answer is probably quite accurate. But, if we used the formula to work out the number of bacteria after six days, we would end up with a mass over 3,000 times that of the Earth. After one week, the bacteria would weigh more than a thousand Suns. In reality of course, food and space would run out within a very few days, and bacterial growth would halt. Fortunately for us!

> SEE ALSO:
> ▸ Proof, page 16
> ▸ Calculus, page 110

LOGARITHMS AND ANTILOGARITHMS

Diophantus (see page 46) noticed that adding powers of numbers is equivalent to multiplying the numbers themselves. So, for example, 100 x 1,000 = 100,000. Rewriting this as $10^2 + 10^3 = 10^5$ shows that his rule is correct: $2 + 3 = 5$.

Here, 2, 3, and 5 are called the "logarithms to base 10" (or log_{10}) of the numbers 100, 1,000, and 100,000. Logarithms can be fractions, too: 2.30103 is log_{10} of 200, which is to say that $10^{2.30103} = 200$. There is nothing special about base 10 here, and in many cases it is more convenient to use other bases, especially base e. $e^{5.534}$, for example, is about 20, so the logarithm to base e (usually abbreviated ln or log_e) of 20 is about 5.534.

Therefore, if a is the logarithm to base b of a number n, then $b^a = n$. So, for example, the logarithm to base 10 of 100 is 2, because $10^2 = 100$. And, the logarithm to base e of 100 is about 4.605, because $e^{4.605} \approx 2.718^{4.605} \approx 100$.

An antilogarithm is the opposite: if a is the antilogarithm to base b of a number n, then $b^n = a$. So, for example, the antilogarithm to base 10 of 2 is 100, because $10^2 = 100$. And, the antilogarithm to base e of 4.605 is about 100, because $e^{4.605} \approx 2.718^{4.605} \approx 100$.

The Fundamental Theorem of Algebra

ALGEBRA IS ABOUT SOLVING PUZZLES, AND IN THE 19TH CENTURY one of the greatest minds in mathematics proved how particular algebra puzzles always have an answer—or two.

Carl Friedrich Gauss, known as the Prince of Mathematics, was the driving force behind the fundamental theorem.

There's no doubt that polynomial equations matter in the history of mathematics. Solving them, or showing why they can't be solved, not only occupied many of the greatest minds since mathematics began, it also led to a whole range of new mathematical concepts, including negative numbers, imaginary numbers, and groups (see page 140). They have a great many scientific applications, too (see box, page 133).

Always an answer

Before setting out to solve a polynomial, it would clearly be very useful to know whether it actually has a solution, and that is what the fundamental theorem of algebra guarantees:

EVERY POLYNOMIAL OF DEGREE n HAS n ROOTS.

The "root" means answer or solution, and is also sometimes referred to a "zero" or "x-value." These roots may, however, be complex. So, cubic equations (which have x^3 as their highest power) have exactly three roots, quadratics have at least two roots, and so on.

To explore this idea of complex roots, let's consider quadratic equations ($ax^2+bx+c=0$), which can be solved very simply, just by using the formula $x=(-b\pm\sqrt{b^2-4ac})/2a$. In this formula, the term inside the square root, b^2-4ac, is called the discriminant, and it tells us what kinds of roots there are. If $b^2 > 4ac$ there are two different roots, if $b^2 = 4ac$ there are two equal real roots, and if $b^2 < 4ac$, there are two complex roots. Let's take these three quadratics as examples:

I. $x^2 + 3x + 2 = 0$
II. $x^2 + 2x + 1 = 0$
III. $x^2 + 3 + 2.5 = 0$

For quadratic I: $a=1$, $b=3$, $c=2$, so the two roots are $x = (-3 \pm \sqrt{(9-8)})/2 = -1.5 \pm 0.5$

For quadratic II: $a=1$, $b=2$, $c=1$, so the two roots have the same value: $x = (-2 \pm \sqrt{(4-4)})/2 = -1$

For quadratic III: $a=1$, $b=3$, $c=2.5$, so the roots are $x = (-3\pm\sqrt{(9-10)})/2 = -1.5 \pm \sqrt{(-1)}/2$

There is no real number which, if multiplied by itself will give -1. Instead, we just use a symbol for "the square root of -1," which is i (see page 78). So, we can write $-1.5\pm\sqrt{(-1)}/2$ as $-1.5 \pm 0.5i$. Therefore, in quadratic III the two roots of the quadratic are $-1.5+0.5i$ and $-1.5-0.5i$.

These are known as complex numbers, meaning they each have two parts, the real number -1.5 on the left and the imaginary numbers $0.5i$ or $-0.5i$ on the right. Thanks to Descartes (see page 92) we know that an alternative way to find (at least roughly) the real roots of a quadratic equation is to plot them (see below):

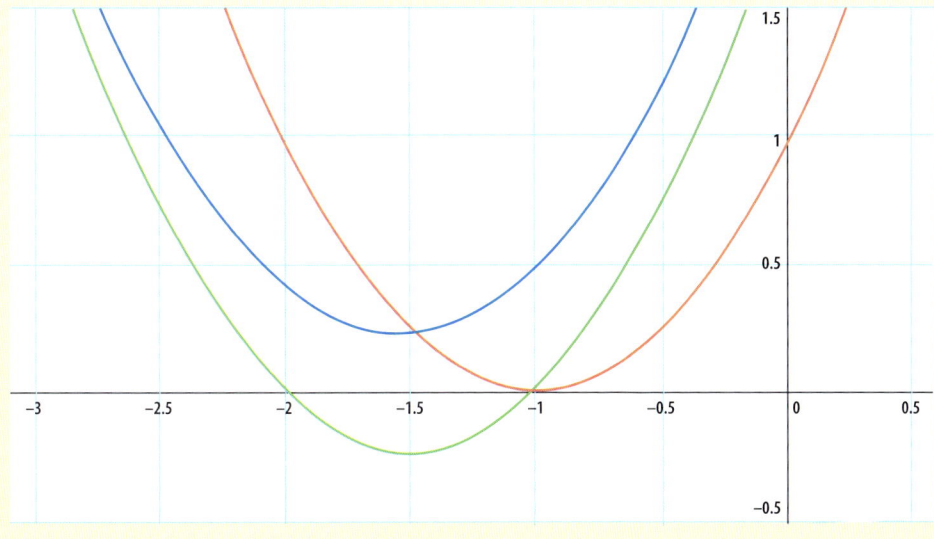

Our quadratic examples I (green), II (red), and III (blue).

For the first two quadratics, we can see the roots clearly. The green line crosses the *x*-axis at the points −1.5 ± 0.5 (−2 and −1), and the red line touches the axis at −1. And the blue line? With a little conjuring, we can make its graph reveal its roots, too. We start with its plot as before:

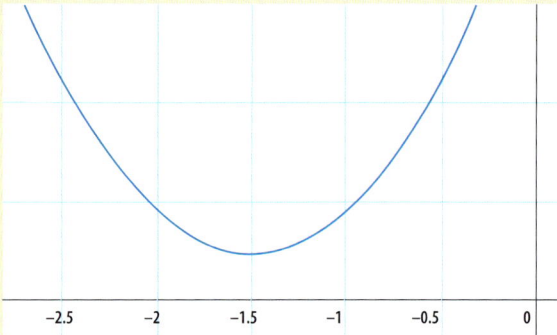

Now, we plot a mirror-curve, a curve identical to the blue one, but upside-down:

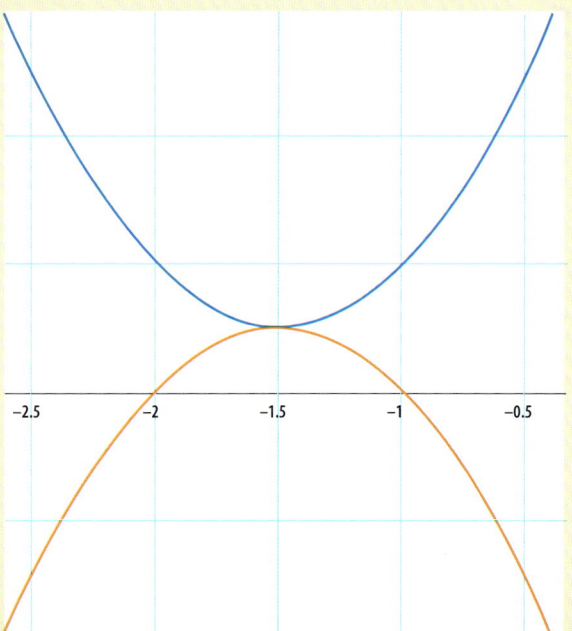

Next, we draw a circle that passes through the points (marked with red dots) where the mirror curve crosses the *x*-axis:

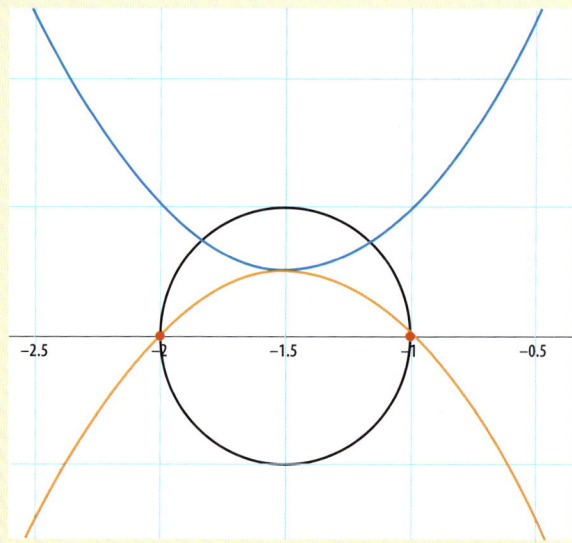

And finally, we imagine rotating the circle through a right angle (effectively changing from the *x*-axis to the *y*-axis), and read off the *y*-coordinates of the red dots. These are ±0.5, which are the multiples of *i* in the solution to this quadratic. The box on page 135 gives a partial explanation of what is happening here.

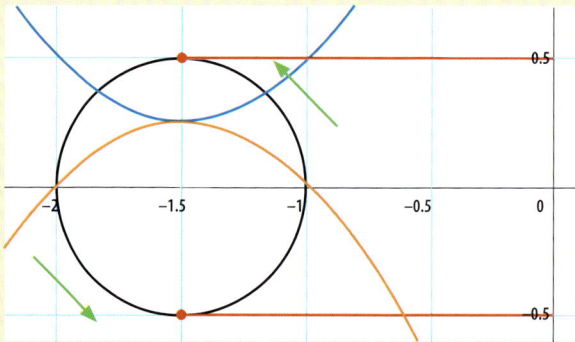

POLYNOMIALS IN ACTION

Quadratic equations define the profiles of telescope mirrors and satellite dishes. A quadratic equation, called the Bernoulli equation, explains how planes fly, and others are used to predict the curving paths of missiles. They are used in many areas of physics, and in chemistry, biology, and economics, too.

If you twist a rubber band, it will wind itself up neatly at first, but then suddenly kink. To get rid of the kink, you need to untwist it further than you did to kink it. There are many other examples of this type of behavior in physics and engineering, and they are described by cubic equations. Quartic equations describe electrical interference effects from motors and other equipment.

Most objects in the Solar System are in orbit around others: The Earth orbits the Sun, the Moon orbits the Earth, and some satellites orbit the Moon. This is because, usually, the only way for an object in space to avoid being dragged down by the gravity of the nearest planet or other large body is to orbit that body. But there are a few points close to planets and moons where objects can remain motionless almost indefinitely. These are called Lagrangian points, and can best be described by a quintic (or power of 5) equation.

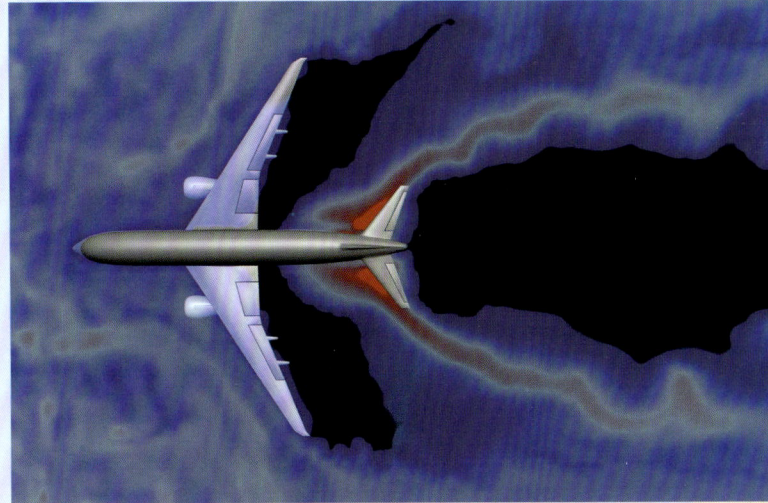

Pauper and prince

Many mathematicians have managed to prove the fundamental theorem of algebra, but the master of this topic was Carl Friedrich Gauss, who found no fewer than four ways to prove it. Along with Archimedes, Newton, and Euler, Gauss was one of the greatest mathematicians of all: Like them, he made important contributions to several areas of mathematics. Also like them, he was an expert in other fields, including engineering, astronomy, and languages. He had so many ideas that he kept a notebook to jot them down, though he wrote them in such an abbreviated form that all 146 of the entries are squeezed into 19 pages. For instance, his theorem that states (correctly) that every integer is the sum of at most three triangular numbers, is recorded as simply as num=$\Delta+\Delta+\Delta$.

Searching for Ceres

Gauss's family was not well-off, but his mathematical brilliance came to the attention of Ferdinand, Duke of Brunswick, who paid for him to go to a much better school. Gauss was 15 at the time, and the duke continued to support him financially whenever he needed it until Ferdinand's death 14 years later.

The dwarf planet Ceres and its discoverer Giuseppe Piazzi.

By this time, Gauss was well known internationally, and perhaps the project that brought him most fame was the discovery of a lost planet. Kepler (see page 86) and other astronomers had always been puzzled by the arrangement of the planets, and in particular by the large gap between Mars and Jupiter. So, in 1801, everyone was very excited, when, on January 1, an Italian monk called Giuseppe Piazzi found a dim world in that mysterious gap. For 41 nights, Piazzi watched the new world (soon named Ceres), and tracked its location compared to the stars. But then he became ill and had to stop—and soon after that Ceres passed behind the Sun and could no longer be seen. Now there was a problem: such a faint world could only be spotted if astronomers knew precisely where to look, and that required knowledge of its orbit. But those 41 days of observations covered only a tiny part of that orbit: 2.4 percent, to be precise.

Finding the curve

Thanks to Kepler and Newton, the laws of the motion of planets were well understood and so

Piazzi's records of the path of Ceres through the sky did not give enough information.

1801	Mittlere sonnen-Zeit	Gerade Aufstig in Zeit	Gerade Aufsteigung in Graden	Nördl. Abweich.	Geocentr. sche Länge	Geocentr. Breite	Ort der Sonne + 20" Aberration	Logar. d. Distanz ☉ ☿
	St ′ ″	St ′ ″	° ′ ″	° ′ ″	Z ° ′ ″	° ′ ″	Z ° ′ ″	
Jan. 1	8 43 17,8	3 27 21,25	51 47 48,8	15 37 43,5	1 23 22 58,3	6 42,1	9 11 1 30,9	9,9926156
2	8 39 4,6	3 26 53,8	51 43 27,8	15 41 5,5	1 23 19 44,3	3 2 24,9	9 12 2 28,6	9,9926317
3	8 34 53,3	3 26 18,4	51 39 36,0	15 44 31,6	1 23 16 58,6	2 58 9,9	9 13 3 16,6	9,9926324
4	8 30 42,1	3 26 23 15	51 35 47,3	15 47 57,6	1 23 14 15,5	2 53 55,6	9 14 4 14,9	9,9926418
10	8 6 15,8	3 25 32,1	51 23 1,5	16 10 32,0	1 23 7 59,1	2 29 0,6	9 20 10 17,5	9,9927641
11	8 2 17,5	3 25 29,73	51 22 26,0	. . . . . .	. . . . . .	. . . . . .	. . . . . .	. . . . . .
13	7 54 26,2	3 25 30,30	51 23 34,5	16 22 49,5	1 23 10 27,6	2 16 59,7	9 23 12 13,8	9,9928490
14	7 50 31,7	3 25 31,72	51 23 55,8	16 27 5,7	1 23 12 1,2	2 12 56,7	9 24 14 13,5	9,9928809
17	. . . . . .	. . . . . .	. . . . . .	16 40 13,0	. . . . . .	. . . . . .	. . . . . .	. . . . . .
18	7 35 11,3	3 25 55,	51 28 45,0	. . . . . .	. . . . . .	. . . . . .	. . . . . .	. . . . . .
19	7 31 28,5	3 26 8,15	51 32 2/3	16 49 16,1	1 23 25 59,2	1 53 38,2	9 29 19 53,8	9,9930607
21	7 24 2,7	3 26 34,27	51 38 34,1	16 58 35,9	1 23 34 21,3	1 45 6,0	10 1 20 40,3	9,9931434
22	7 20 21,7	3 26 49,42	51 42 21,5	17 3 18,5	1 23 39 1,8	1 41 28,1	10 2 21 32,0	9,9931886

the general shape of the orbit (an ellipse) was known. If Piazzi's observations were accurate enough, calculating the ellipse would have been simple. But in 1801, such precision was just impossible with the instruments available.

When plotted, Piazzi's data could be fitted to so many different orbits that it ended up being useless at predicting the position of Ceres when it emerged from behind the Sun: All the different curves covered too huge an area of the sky. One of the greatest mathematicians of the day, who was also an astronomer, was Pierre-Simon Laplace, and he announced that the problem could not be solved. But Gauss managed it: In November 1801, he completed and released his predictions of the location of Ceres, and in early December what looked like Ceres was spotted very close to the position he had predicted. It was necessary to wait a short time so that astronomers could confirm that the object was moving as predicted (otherwise it might turn out to be another object that just happened to lie in that direction). On December 31, almost exactly one year after its original discovery by Piazzi, the observations were confirmed. Ceres had been found again (and is now known to be the nearest dwarf planet). Gauss became internationally famous, and Laplace said he was "a super-terrestrial spirit in a human body."

SEE ALSO:
▶ Algebra Moves East, page 60
▶ The Rules of Algebra, page 82

SHIFTING DIMENSIONS

As mentioned on page 81, we can plot complex numbers on a graph, called an Argand diagram, in which the x-axis represents the real part of the number, and the y-axis represents the imaginary part. In fact, this way of treating complex numbers as pairs of coordinates suggests that complex numbers can be thought of as two-dimensional things, as opposed to the one-dimensional real numbers we are more used to.

This is the Argand diagram of the solutions of the three equations on page 131:

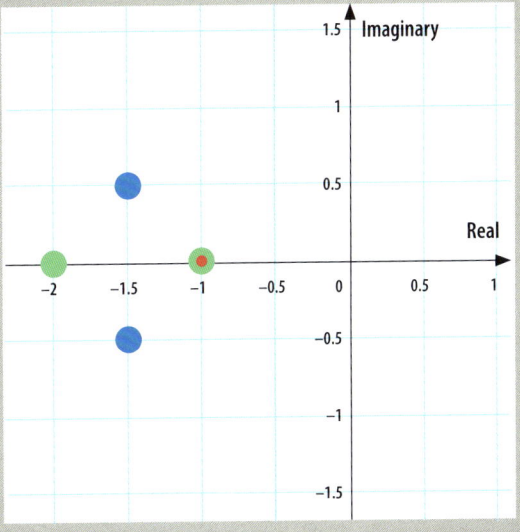

So what we did in the main text in rotating the positions of the solutions was to treat the y-axis as if it was an imaginary axis, just for those particular points.

The Fundamental Theorem of Calculus

AFTER NAPOLEON BONAPARTE BECAME EMPEROR of France in 1804, he strongly encouraged the use of mathematics—including calculus—for practical purposes.

Augustin-Louis Cauchy started out as an engineer, working on the Napoleon Dock at Cherbourg, which was finally finished the year after his death.

Students prepare a giant battery at the École Polytechnique in Paris in 1813. The battery was another invention championed by Napoleon Bonaparte.

One young mathematician who responded to this imperial call was Augustin-Louis Cauchy, who attended the École Centrale du Panthéon from 1802. It was renowned as the best school for budding engineers, offering training in math and science, together with their applications.

Following great success there, Cauchy continued his education as an engineer at the School for Bridges and Roads, following which he was offered a job as an engineer at Cherbourg, where Napoleon planned to build a powerful naval base. However, Cauchy gradually lost his interest in engineering and became attracted instead to pure mathematics, and when he became ill from overwork in 1812, he moved to Paris. He was 23.

Strange character

It may have been comforting for Cauchy to live in the non-human, calm, and perfect world of mathematics, because he was strongly disliked by many of his colleagues. Extremely religious, very right-wing, and openly contemptuous of many other mathematicians, he seems to have been a difficult man to like. His need for certainty, stability, and order may have originated in his childhood. He was born just after his parents had fled Paris to escape the dangers and horrors of the French Revolution, and he grew up listening to the frightening stories they told.

Calculus was also in need of some certainty just then. It had always been used primarily for practical purposes ever since it was formulated by Newton and Leibniz in the 17th century (see page 110). For the next few decades, mathematicians and scientists focused on its great power to solve problems in many fields. But, by the late 18th century, there was a new feeling in the air. Mathematics was now such a powerful and useful subject, it was becoming increasingly important that it be based on rigorous proofs. The standard of mathematical proof at the time was very variable, and it was still a matter of opinion, skill, and taste as to whether the availability of water-tight proofs was even important. After all, in many subjects, from biology to economics, it is rarely possible to prove anything for certain, and yet those subjects are highly successful. On the other hand, during the great period of ancient Greek mathematics, rigorous proof was the main goal.

Older techniques

It was the ancient Greeks who posed the questions (and found basic solutions for) the problems that calculus would one day answer. These included calculating the slopes of tangents to curves, and working out the areas under lines (see page 41). By the late 17th century, it was known that tangents to curves can be found by differentiation, and areas under lines by integration.

That is: If we have an expression for the area under the line, we can differentiate it to get the equation of the line itself, or, if we have an expression for the slope of the line, we can integrate it to get to the line again.

Opposites?

So, it seems obvious that in some way differentiation and integration are opposite.

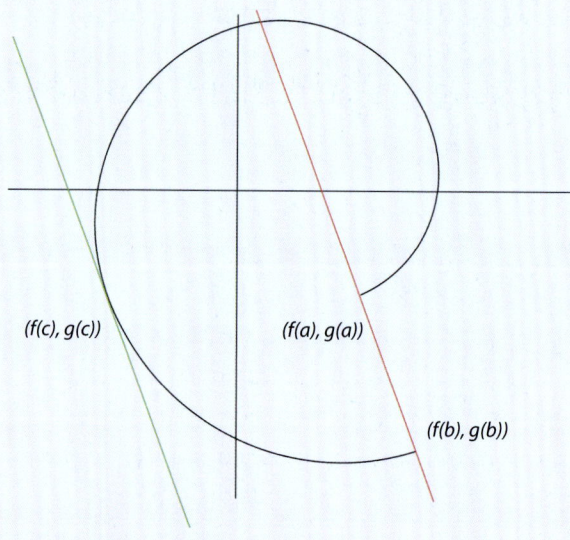

Only Leonhard Euler produced more writings on mathematics than Augustin-Louis Cauchy. In total, Cauchy's work fills 27 volumes, and includes the mean value theorem, illustrated above, which bears his name.

However, things in math and science that "seem obvious" have a way of coming back and biting you (see page 16).

Here's an example. If a small amount of water is placed in a sealed container and heated, the water will turn to high-pressure steam and the container will explode. On the other hand, if a similar container is filled with steam and sealed, it will implode, becoming crushed. These seem simply like opposite processes, as if the steam generates an outward force as it forms, and an inward force as it condenses. Two kinds of force, equal and opposite: an outward-pushing force when steam forms and a pulling-inwards force when it disappears.

But in fact, these are not opposites at all, as would be demonstrated if the experiment were to be repeated in outer space. The heated vessel would still explode, but the cooled one would not collapse. In the first case it is the pressure generated by the expanding steam that causes the explosion, but in the second case the steam doesn't produce anything at all. Instead, as it condenses, it leaves a partial vacuum behind, and the crushing force comes not from the water or steam or anything else inside the vessel, but only from the pressure of the air outside on the vacuum. If there is no air outside, as in space, there is no crushing.

The final theorem

So, some mathematicians began to look for a proof that differentiating one function to get a

A process that works one way will not always work the opposite way—even in mathematics.

second one really is the converse of integrating that second function to get back to the first one again. It was Cauchy who found this proof, as just a part of his grand project of clarifying mathematics, extending it, applying it wherever possible, and placing it on a firm theoretical footing.

This proof is known as the fundamental theorem of calculus.

> SEE ALSO:
> ▶ The Prehistory of Calculus, page 40
> ▶ Calculus, page 110

Groups

FOR MANY CENTURIES, MATHEMATICIANS STRUGGLED WITH QUINTIC EQUATIONS. Most had no solutions, and no one knew why. Then a brilliant Frenchman discovered the reasons—and opened up a new area of math.

Évariste Galois just made it out of his teens before he died, but still managed to create a new branch of mathematics.

Much of the history of algebra has been the story of the quest for the solutions of polynomial equations: Babylonian mathematicians found out how to solve quadratics ($ax^2+bx+c=0$), and general solutions for cubics ($ax^3+bx^2+cx+d=0$) and quartics ($ax^4+bx^3+cx^2+dx+e=0$) were known by the end of the 16th century. But quintic equations ($ax^5+bx^4+cx^3+dx^2+ex+f=0$) stubbornly resisted solution. Mathematicians knew how to solve some quintics, but no one could find a formula into which they could feed the coefficients (a, b, c, d, e, and f) of any given quintic, to calculate its roots.

Young pretender

The mathematician who found out why was Évariste Galois, and the way he did it opened the door to a new world of mathematics. In part, he did this by building on the work of Cauchy—but the two men could hardly have been more different. Galois was a radical, completely opposed to the traditional system that Cauchy believed in so fervently, of social classes and privilege due to one's rank at birth.

Short life

Galois had a very dramatic and difficult life, failing to get much of his work recognized due to his great difficulty in expressing himself clearly. For example, he wasn't admitted to the leading University in Paris, the École Polytechnique, and had to make do with what was then the far inferior École Normale Supérieure. But he was expelled from there when he wrote a letter to a newspaper criticizing the school's director. Then, after the king disbanded the French National Guard because he believed they were likely to revolt, Galois wore a Guard uniform in public. This very provocative act landed him in jail. There, he fell in love with Stéphanie-Felice du Motel, the daughter of the prison's doctor. Soon after his release, and for reasons that are unclear

Galois lived though the July Revolution, a deadly insurrection in France in 1830.

Dueling was illegal in France in the 1830s, but it was a common way for young men to settle scores.

but are probably connected with Stéphanie, he was killed in a duel. He was only 20.

Life's work

Just before the fatal duel, Galois wrote out some of his discoveries and left them for a friend to publish. One of them was a new approach to tackling the quintic, not by looking for a formula to solve it, but by investigating whether such a formula was even possible. Galois was not the first to try this, but his approach was far more general than that of any of his predecessors.

Rather than looking at specific quintics, or quintics in general, Galois adopted an even more abstract approach, by studying the polynomials in general. This was a very challenging idea, because it means tackling this, an equation for which we can't even state the number of terms:

$ax^n + bx^{(n-1)} + \cdots + cx + d = 0$

If Galois could unlock the secrets of this equation, he would be able to determine the solvability of any polynomial of any degree at all. "Degree" means the value of the exponent of the highest power in the equation, so, for instance, any quadratic has a degree of 2, because x^2 is its highest power.

Using symmetry

As with any area of math, the more abstract a problem becomes, the harder it is to even start to solve it. Galois found that the answer lay in the concept of symmetry. Symmetry is a very important idea in art and in geometry, too, but it's not obvious how to turn information about symmetry into numbers and formulas. Actually, it's not even very easy to state in words exactly what symmetry means, though it's easy enough to recognize a symmetric object. In fact, the best way to define symmetry is to say that an object is symmetrical if some change can be made to it that leaves it looking the same as it was to start with. Take this equilateral triangle. Is it symmetrical?

Much of what we find beautiful is due to the multiple symmetries we see in objects.

If we drew a dotted line down the center, and placed a mirror on that line, the reflection would look just like the other half. So it is symmetrical, because it looks just the same, whether it has been reflected or not. This is called reflection symmetry.

We can reflect the triangle along any of three axes, so it has three axes of mirror symmetry.

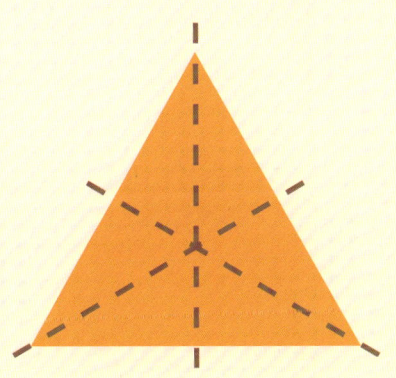

Does the triangle have any other symmetries? Yes, it can be rotated by 120°, 240°, or 360°, and in each case its shape will end up unchanged. (In fact, rotating by 360° is effectively doing nothing at all).

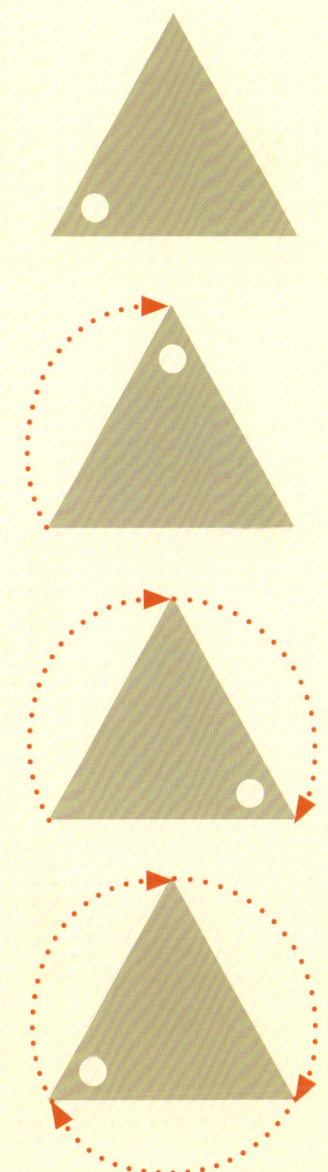

These are the only symmetries the equilateral triangle has—the only ways we can change it which leave it looking just as it did to begin with. We call these six symmetries the symmetry group of the equilateral triangle.

The study of symmetry is called group theory, and many mathematicians contributed to it before Galois, especially Cauchy. But in applying group theory to polynomials in order to investigate their solvability, Galois was inventing an new area of math that is now called Galois theory. He was able to show that, while polynomials of degrees 4 and lower possessed certain symmetries, these symmetries were not present for higher degrees. This meant he could prove that a general solution was impossible for quintics, and for any higher-order polynomials, too.

Permutations

But what does "symmetry" mean in terms of equations? Unlike triangles, the symmetries of an equation can't be found just by looking at it (which makes sense, since the same equation can often be written in many ways). And, to investigate equations, we don't reflect or rotate them. Instead, the changes we make are called permutations.

Permuting an equation means swapping round its x and y terms. So, permuting $x = y + 1$ gives $y = x + 1$. What does this permutation do the equation? We can find out by plotting the original version and the permuted version onto the graphs shown on page 146.

GROUPS

One morning at 10 o'clock you're just setting off on a 4-hour journey. What time will you arrive? Getting to the right answer involves something called modulo-12 arithmetic. In ordinary arithmetic, $10 + 4 = 14$ but in modulo-12, $10 + 4 = 2$, so the answer is 2 o'clock. While ordinary arithmetic can be thought of as moving along a number line, modulos are more like an arithmetic cycle, so with every multiple of 12 (in this case) the value resets to 0. So modulo-12 can be thought of as moving around a clock face:

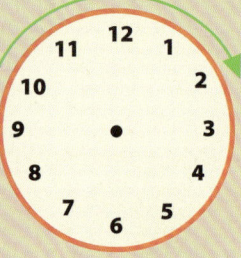

Not all clocks have 12 numbers on their faces. Some of the earliest public clocks in 14th-century Italy had 24, and during the French Revolution, when there were plans to change the numbers of hours in a day, some clocks had 10 hours on their faces. Each of these has its own modulo. In modulo-24, $10 + 4 = 14$, in modulo-10, $10 + 4 = 4$.

One major difference between modulo arithmetic and the ordinary kind is that, while there are endless different answers, sums, and results in ordinary arithmetic, modulo arithmetic includes only a few. For example, this table shows all the possible numbers that can be used (or can result) if we add numbers according to modulo-3. (It is called a Cayley table.)

+	0	1	2
0	0	1	2
1	1	2	0
2	2	0	1

So, for instance, adding 2 and 2 in modulo-3 takes us back to 1:

+	0	1	②
0	0	1	2
1	1	2	0
2	②	0	①

We can construct tables like this for any group. For instance, if we list the various possible ways and outcomes of rotating an equilateral triangle so that it ends up looking the same (that is, listing the rotational symmetries of the triangle) we get:

R	0°	120°	240°
0°	0°	120°	240°
120°	120°	240°	0°
240°	240°	0°	120°

There is no "rotate by 360°" entry because that is exactly the same as doing nothing at all.

Since everything in the table is either zero, or a multiple of 120°, we can simplify the table by just listing how many lots of 120° rotations there are:

+	0	1	2
0	0	1	2
1	1	2	0
2	2	0	1

And this is exactly the same as the modulo-3 addition table. So, the same group describes both the triangle rotations and the modulo-3 additions.

This group is so small and simple that this result is not particularly exciting, but more complicated groups don't just link up different areas of math, they link up math with the real world.

For instance, subatomic particles also have various kinds of symmetry—that is, there are changes which can be made to them which leave them ending up the same as before the change was made.

In 1962, scientists were trying find some kind of natural classification system for the many subatomic particles they knew about. Murray Gell-Mann used group theory to help with this and constructed complicated Cayley tables. But there were strange gaps in those tables. Scientists tried looking for a particle with just the right symmetries to fill those gaps, and found one in 1964, the Ω– ("Omega minus") particle.

That same year, a number of scientists helped to predict another particle by using group theory. And the very same thing happened again in the 2000s, when scientists realized that, if only there was just one more kind of particle in the Universe, the groups that described the symmetries of the particles they knew would be simpler. They looked for a particle with just those symmetries, and found the Higgs boson.

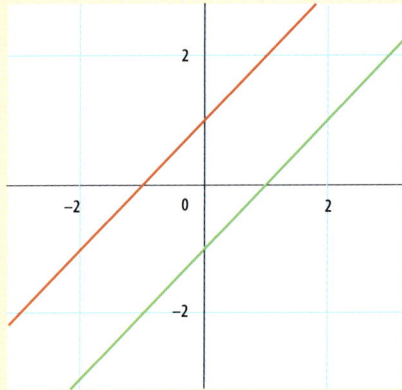

The lines are different, so this particular permutation has not found us a symmetry. But, if we try permuting $x = 1 - y$ to give $y = 1 - x$, and plot these two equations, we find that they give the same line:

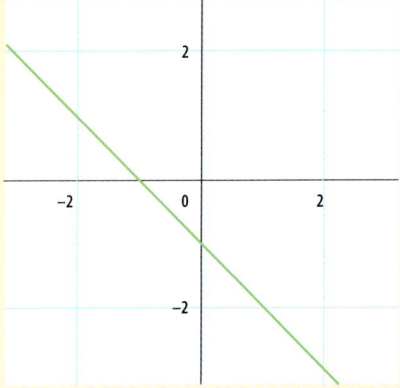

So, this permutation is a symmetry. Just as all the symmetries of an equilateral triangle make up a symmetry group, the permutations of an equation make up a permutation group.

> **SEE ALSO:**
> ▶ Equations, page 46
> ▶ Abstract Algebra, page 158

PARTICLE SYMMETRIES

Sub-atomic particles are too small to see, and any attempt to measure them changes their characteristics. What can be seen and measured, however, is what happens when sub-atomic particles change. Some particles change all by themselves, while others may be changed by a collision with another particle.

Like shapes or equations, particles can be studied through their symmetries: If a change is made to a particle, but the particle continues to behave as it did before the change was made, that means we have identified a symmetry.

Imagine that a pair of electrically charged particles collides and produces a pair of new particles and a flash of violet light. Now, let's introduce some change, such as reversing the electrical charges of the first two particles. Next, the experiment is repeated. If we get the same result as before (two particles and violet light), that means that swapping the electrical charges has had no effect, so we have identified a symmetry. So we can say that the particle interaction is symmetrical with respect to charge reversal. (This is charge conjugation symmetry.)

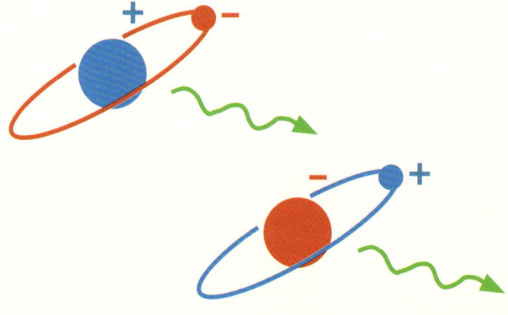

Another kind of particle symmetry involves spin. Most particles spin, either clockwise or anticlockwise, and "parity" symmetry means that particles continue to behave in the same way if their spins are reversed. This can be thought of as a kind of reflection symmetry, because the spins of particles would appear reversed if they were seen in a mirror.

Time reversal symmetry means that, if particles interact in some way, then that interaction would reverse if time were to go backward. This hardly seems worth saying—after all, if we imagine that time goes backward, isn't it obvious that everything would reverse, just like a film which is put into a projector the wrong way round, and in which we see people falling upstairs, eggs unbreaking, explosions turning into bombs, and so on? In fact, isn't that what we mean by "time going backward"?

What makes it worth saying, though, is that there are examples of particle interactions involving K and B mesons which, it is believed, would not reverse if time went backward.

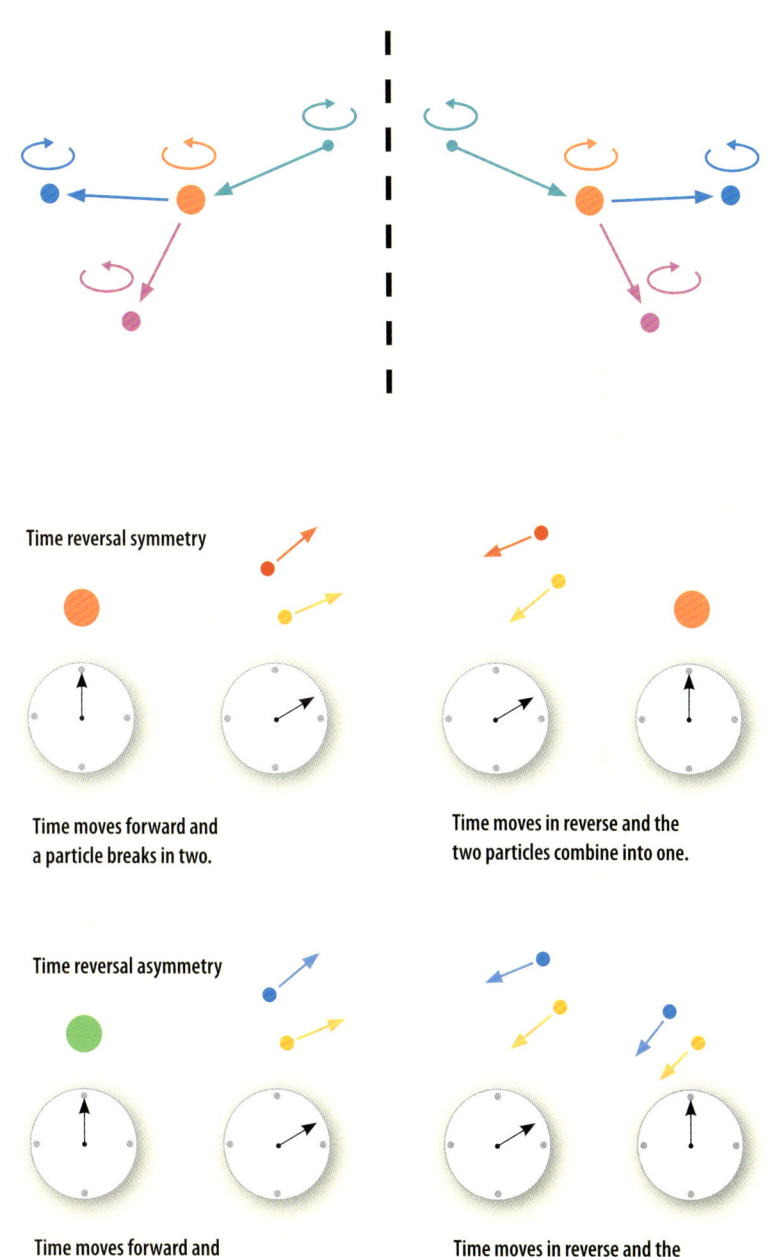

Time reversal symmetry

Time moves forward and a particle breaks in two.

Time moves in reverse and the two particles combine into one.

Time reversal asymmetry

Time moves forward and a particle breaks in two.

Time moves in reverse and the two particles stay separate.

Quaternions

MANY SMARTPHONE APPS NEED TO KNOW THE ORIENTATION OF THE PHONE, and pilots need to know the orientation of their planes. Both jobs are done by quaternions.

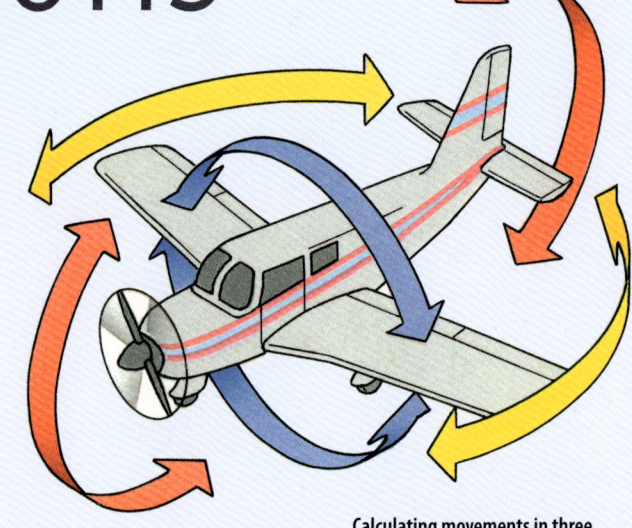

Calculating movements in three dimensions requires a four-dimensional coordinate system.

To see how quaternions work, we must return to the way complex numbers can be plotted on an Argand diagram (see page 135), like this. Here, the complex number being plotted is $(1 + 1i)$.

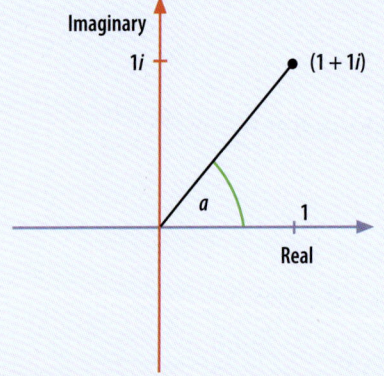

The angle (a) here is 45°, and we can generate several multiples of that angle just by plotting other complex numbers, see top right, page 149. We can easily work out from this how to represent rotations. There is a 90° angle between the plots above and $(-1 + 1i)$. $(1 + 1i)$ can be transformed into $(-1 + 1i)$ by multiplying by i: $i \times (1+1i) = (1i+1ii) = (1i + 1i^2)$. Since $i^2 = -1$, this becomes $1i + (1 \times -1))$, which is $(1i, -1)$.

Although this is written "backwards" compared to what we are looking for, which is $(-1+1i)$, it is still plotted in the same way: the real part is plotted along the horizontal axis, and the imaginary part up the vertical. So, what that all boils down to is that to rotate by 90°, just multiply by i. We can rotate to any position we want by multiplying by i, or by some fraction or multiple of i, (see box, opposite).

New dimensions

Of course, real objects like planes and phones aren't constrained to just two dimensions. To keep track of their orientations, we can use quaternions, which were invented by William Rowan Hamilton, an Irish mathematician,

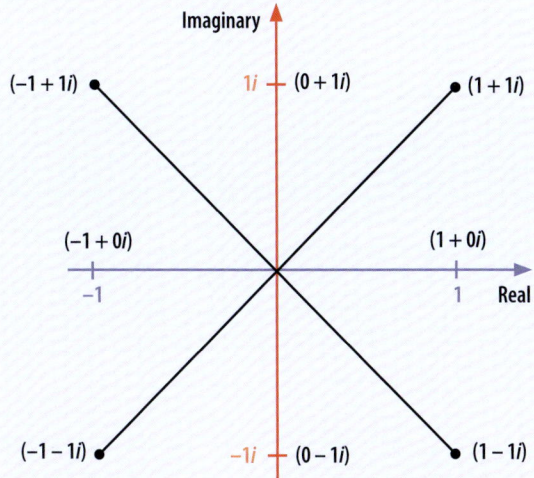

in 1843. It took him 15 years (on and off) of research to develop them, so it was just as well he loved math so much.

Hamilton was interested primarily in extending the idea of complex numbers. Since a complex number ($a + ib$) is made up of two parts, one real (a), and one imaginary (ib), Hamilton's idea of a "hypercomplex number" was to add a third part, to give ($a + ib + jc$). Hamilton decided that j, like i, could best be understood as a square root of -1. However, no matter what he tried, Hamilton could not make his "triples," as he called them, work. In particular, he could find no way to multiply them together to give an answer that made sense. His children got so used to his seemingly unending quest that they used to greet him every morning by saying, "Papa, can you multiply triples?" And every morning he had to say no.

The answer, when it finally came to him on a walk with his wife, was so sudden, unexpected, and welcome, that he rushed over to the nearest

ROTATION IN A CIRCLE

Although we can rotate through different angles just by multiplying by i, the point we plot isn't moving in a circle. To make it do that, we will need to keep constant the distance of the point from the origin (the point (0, 0i), where the axes cross). In other words, the green line needs to be 1 unit long throughout.

The line to the point representing (1+1i) can be seen as the hypotenuse of a right-angled triangle, in which the other two sides have the same length (this is called an isosceles triangle).

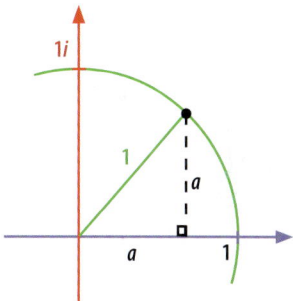

So, from Pythagoras's theorem we have $a^2 + a^2 = 1^2$. So, $2a^2 = 1$, which means
$$a = \frac{1}{\sqrt{2}}$$

And this means that the complex number represented by the point we want is at
$$\left(\frac{1}{\sqrt{2}} + \frac{1i}{\sqrt{2}}\right)$$

bridge and scratched the answer on it with his pocket knife. This is what he wrote:

$$i^2 = j^2 = k^2 = ijk = -1$$

In other words, he needed to introduce two new terms, not just one. If Hamilton had been focused on the challenge of representing rotations in three-dimensional space, he would probably have made his breakthrough many years earlier. However he only realized that quaternions could represent rotations after he had formulated them.

Twisting in space

To define an axis of rotation, we need two numbers, such as the values of the angles ϑ and φ, as seen top right. Then, to describe the point that is being rotated, we need to know its distance (δ) from the origin, and its distance from the axis of rotation (shown here as the angle γ). So, we need four numbers to define a rotation like this. The reason we only need two numbers to define a rotation in two dimensions is that we don't need to worry about the axis of that rotation, since it never changes. Every quaternion

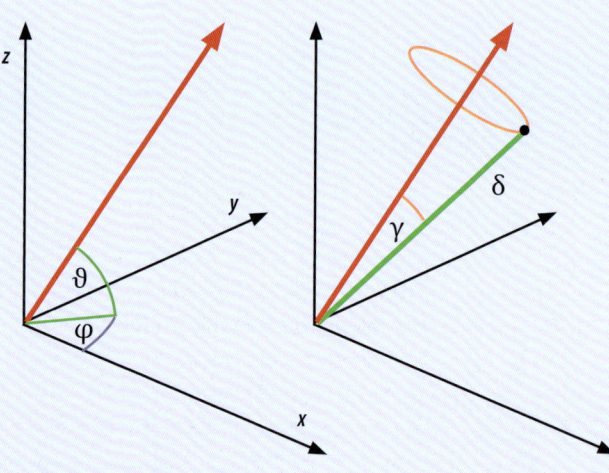

has the form $(a + ib + jc + kd)$, where a, b, c, and d are real numbers and i, j, and k are roots of -1. The reason that Hamilton's breakthrough took so long is the same thing that makes it a breakthrough. Quaternions only work if these rules apply:

$i \times j = k$ $j \times i = -k$
$j \times k = i$ $k \times j = -i$
$k \times i = j$ $i \times k = -j$

Obvious yet false

This idea must have been almost as hard for Hamilton to accept as the idea of imaginary numbers themselves had been to Viète (see page 82). A rule of arithmetic which is so basic that we don't really think about it is that $xy = yx$. So, $2 \times 3 = 3 \times 2 = 6$. It makes no difference

Hamilton's graffiti on Brougham (Broom) Bridge, Dublin, has long gone, but there is now a plaque to mark the event.

SCALARS AND VECTORS

Whatever a quaternion is used to represent, the first term is always just a number. But the other three terms can be thought of as distances, one in each dimension. Hamilton called the first term a "scalar," and the other three "vectors." It turned out that vectors were highly useful in physics, and they have remained so ever since. Any quantity with a direction is a vector (like the velocity and acceleration of the flares, right) and any quantity without direction is a scalar (like temperature, mass, or density). Vectors can be added, subtracted, multiplied, and divided.

in which order we multiply the numbers, so mutiplication is "commutative." Addition is also commutative ($x + y = y + x$), but subtraction and division are not ($x - y \neq y - x$), ($x/y \neq y/x$). For ordinary numbers, the idea of non-commutativity of multiplication makes no sense. Here are fifteen counters. We don't need to count them, as there are three rows and five columns, and $3 \times 5 = 15$.

Shuffle some of the counters, and we have five rows and three columns, and the total is now given by 3×5. But surely no one would actually bother doing this second calculation; it seems so obvious that the answer must be the same.

To Hamilton, the idea that $i \times j = j \times i$ must have seemed equally obvious. But it was false; in other words, for quaternions, multiplication is not commutative.

This was the first time that such a clear break had been made between numbers and algebraic symbols, and in this sense, it was the first step in the development of abstract algebra (see page 158). Following this step, other mathematicians began to investigate algebraic systems in which the law of commutativity and other similar laws did not hold. While this makes quaternions sound very strange, in fact they were the source of some very practical mathematical tools: vectors (see box, above).

SEE ALSO:
▸ Unreal Numbers, page 74
▸ Algebraic Geometry, page 92

The Mathematics of Thought

LOGIC HAS BEEN STUDIED SINCE THE DAYS OF SOCRATES AND ARISTOTLE. In 1833, an English mathematician decided to describe it in a new way—using mathematics.

One January day in 1833, George Boole was out walking with his wife in Doncaster, the town where he worked as an assistant teacher. He had taken the job to support his parents, brothers, and sister when his father's business had failed, but what he really wanted to do was study mathematics at college level. The family's poverty made that impossible, so he did his best to learn it for himself. The couple were just crossing a frosty field when Boole was struck by the most amazing idea. Mathematics, in the hands of Isaac Newton, had been highly successful in explaining the workings of the physical world. Why

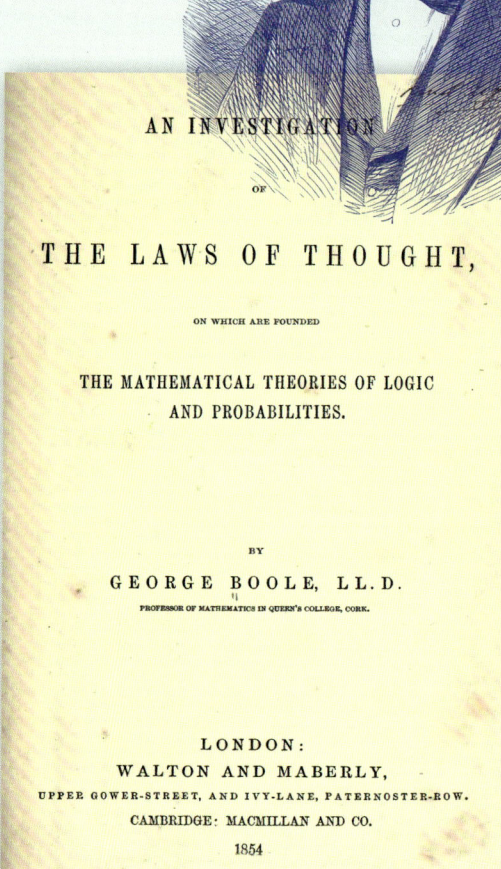

George Boole's book *The Laws of Thought*, published in 1854, was the foundation of information theory and digital computing.

shouldn't it be just as successful at analyzing the workings of the human mind?

Classical thought

Answering that question was to occupy the rest of Boole's life, and the lives of many later scientists and mathematicians as well. As Boole knew, the first steps in this direction had been taken in about 350 BCE by Aristotle, in his work *Prior Analytics*, in which he set out the principles of what would later be called logic. In particular, Aristotle discussed arguments like this:

"ALL MEN ARE MORTAL
SOCRATES IS A MAN
THEREFORE SOCRATES IS MORTAL."

Not a very earth-shattering statement by itself, but what Aristotle saw was that he could extract a pattern of thought from that argument:

All A is B
C is an A
Therefore C is a B

And this pattern can be applied to all arguments of this sort. This is really exactly like algebra:

$a^2+b^2=c^2$ works for:
$a=3$, $b=4$ and $c=5$, and also for
$a=5$, $b=12$ and $c=13$ and
$a=7$, $b=24$ and $c=25$
and so on.

Similarly, Aristotle's argument pattern
All A is B
C is an A
Therefore C is a B
works whether:
A = "Mammals," B is "Have lungs" and C is "Dogs"
A = "Polygons," B = "Symmetrical," C = "Squares."

But Boole's system was more powerful by far than this. He found ways to turn the concepts "and," "or," and "not" into algebraic expressions.

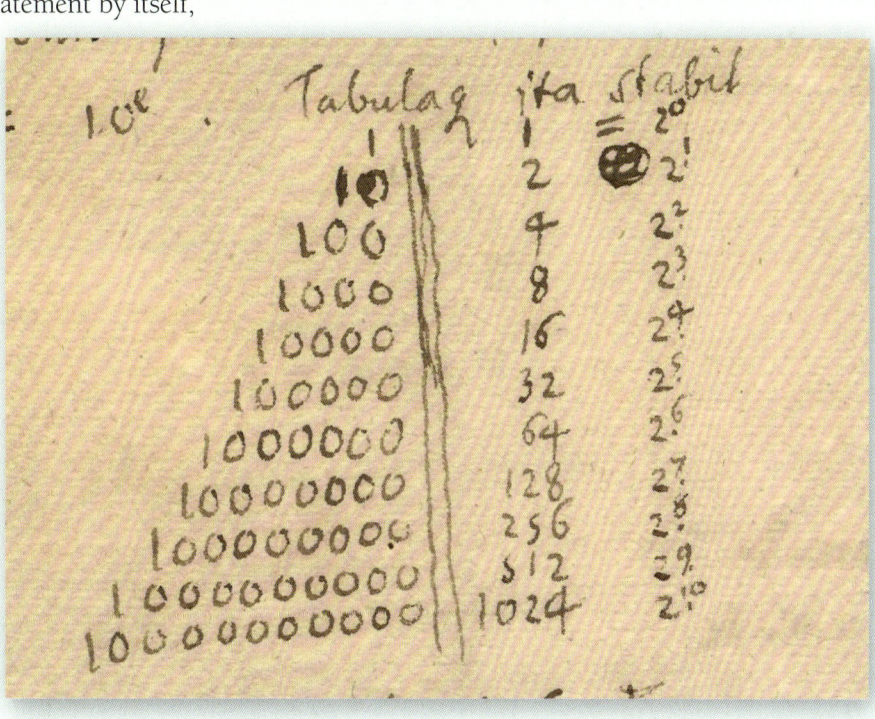

Gottfried Leibniz explored the mathematics of base two, or binary, in the 17th century. Boole also used binary in his algebra.

Finding truth

His first step was to find a way of representing "true" and "false" in terms of numbers. The answer to this had actually been provided nearly two centuries before, by Leibniz. In around 1679, Leibniz devised the binary (or base 2) system. The great advantage of binary is its simplicity; it requires only two numbers, 1 and 0. This was just what Boole needed: a system in which any value could be represented by just two symbols.

So, by adopting binary, Boole was able to convert numbers to "truth values": 1 = "true," 0 = "false." Armed with this idea, Boole and other mathematicians who followed him found ways to represent the basic elements of arguments, including NOT, AND, and OR. These elements are called logic gates, because, like a gate, they can be open or closed. Below is one way of drawing an AND gate.

Here, A and B are called inputs and Y is the output. If A is true AND B is true, then the output, Y, will also be true; it is as if the symbol in the middle is a gate, which opens only when both A and B are true. But if A and B are both false, then the gate will remain closed, and Y will be false.

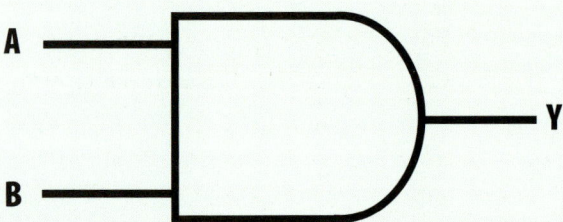

Explaining logic gates in words is clumsy, but truth tables are much clearer. They list all the possible inputs to a gate, and show the corresponding outputs. This is the truth table for the AND gate:

Input A	Input B	Output
False	**False**	**False**
False	**True**	**False**
True	**False**	**False**
True	**True**	**True**

Or, in numbers

Input A	Input B	Output
0	**0**	**0**
0	**1**	**0**
1	**0**	**0**
1	**1**	**1**

Thanks to John Venn, who in 1881 invented the diagrams named after him, AND can also be represented graphically:

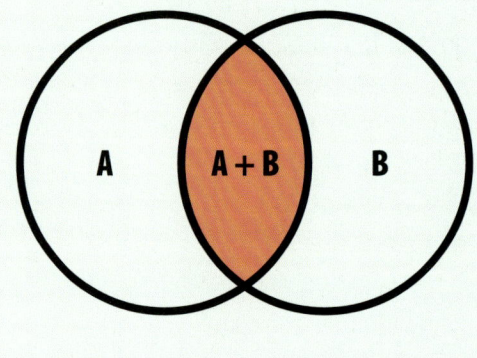

OTHER LOGIC GATES

Other logic gates include OR and NOT. NOT only takes one input, and outputs the contrary. If the input is ON or TRUE or 1, the output is OFF or FALSE or 0, and vice-versa:

OR

Input A	Input B	Output
0	0	0
0	1	1
1	0	1
1	1	1

NOT

Input	Output
0	1
1	0

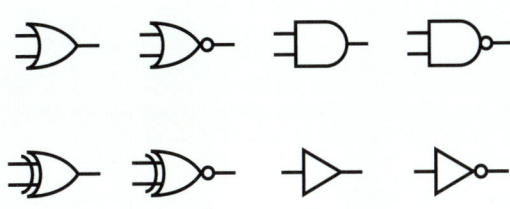

The other logic gates defined by Boolean algebra each have a distinct symbol.

Or, most briefly of all, it can be represented by the symbol "+," as in A + B.

Computer science

Thanks to his mathematical breakthroughs, Boole achieved his ambition of studying math at college in 1849, when he was appointed as the first Professor of Mathematics at Queen's College Cork, in Ireland. Boole's invention was to be the key to the development of the computer, which does exactly what he wanted and uses mathematics to reason. By Boole's time, the computer had already been designed, by Charles Babbage in the 1820s. Babbage's "engines" would have been mechanical, steam-driven, and worked in base 10. But he made little progress in actually building them. Although Babbage thought of his computers as mathematical machines only, his friend and colleague Ada, Countess of Lovelace, who helped to document them, did appreciate that they could work with all kinds of data.

Programmable devices

The first machine controlled by binary inputs had been built even earlier. In 1801, Joseph Marie Jacquard constructed an automatic loom in which the patterns were controlled by cards with holes punched in them. The holes guided colored threads. Each hole represented a 1, each unpunched spot a 0. Babbage's engines would have been controlled by cards like these, and Lovelace referred directly to Jacquard's invention, writing that, "We may say most aptly that the Analytical Engine weaves algebraic patterns just as the Jacquard-loom weaves flowers and leaves."

Looking back now, it's easy to see that mechanical systems would always struggle to automate human reasoning. A different approach was

Right: When a triode valve is switched off, no electricity flows through it. But when the filament is switched on, it heats the gas in the valve, producing ions. These ions allow the gas to conduct electricity, which flows between the grid and the anode.

flow of electricity. An electronic AND gate is built from two triodes. When either or both switches are off, no current can flow around the circuit. When both switches are closed, the filaments

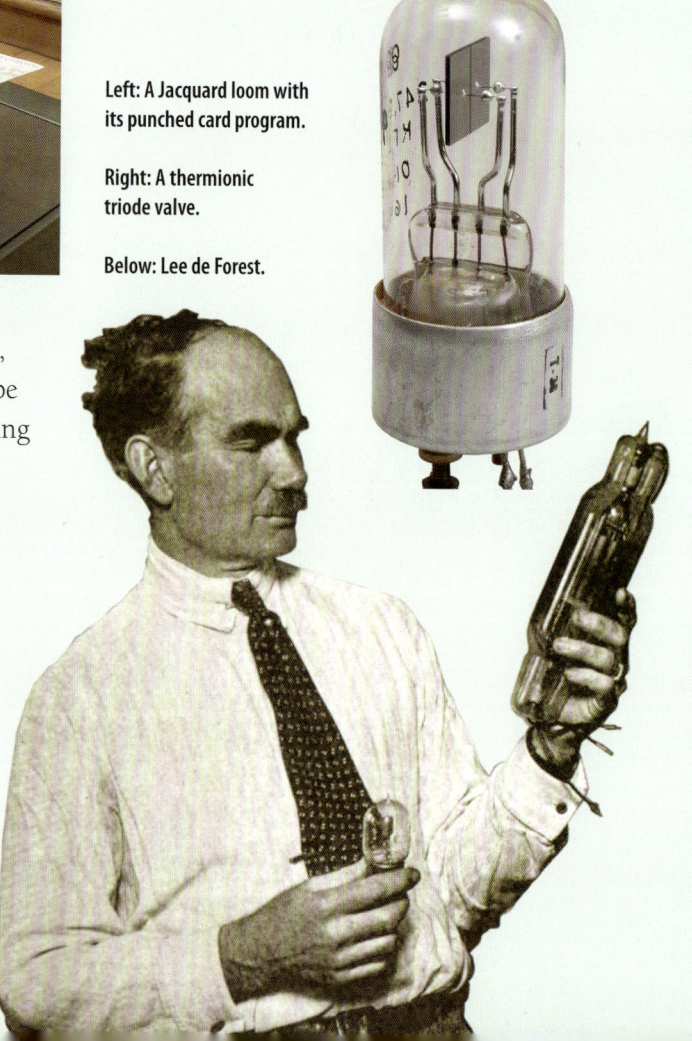

Left: A Jacquard loom with its punched card program.

Right: A thermionic triode valve.

Below: Lee de Forest.

suggested in 1886 by Charles Sanders Peirce, who thought that logic gates could perhaps be built from electrical switches. The first working digital computer did indeed use electrical switches. This was the Z3, built by Conrad Zuse in 1941.

Valves

The real breakthrough in the development of reasoning machines was the invention of the triode valve by Lee de Forest in 1906. The triode was one of the first electronic devices, and it is literally a gate, opening or closing as required to allow or to block the

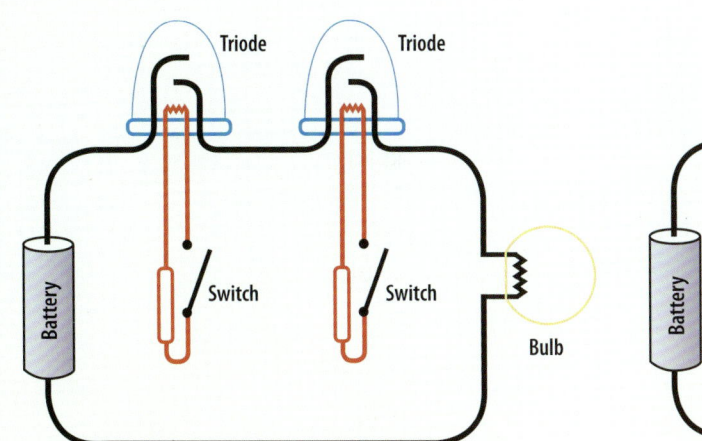
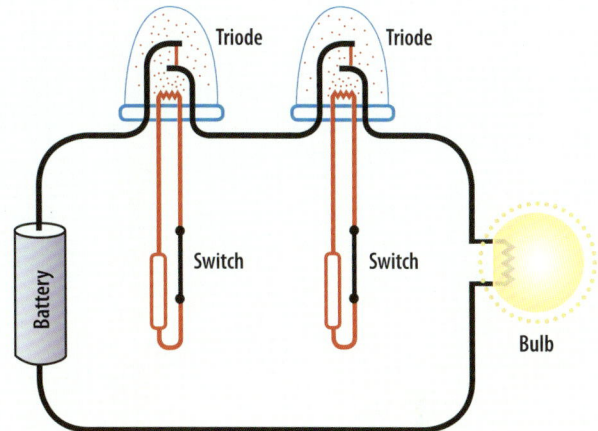

heat up and produce ions in the valves. This allows current to flow through both valves, and therefore round the whole circuit, lighting the bulb. The switches are the two inputs to this AND gate, and the bulb is the output.

Processing information

From 1942, a number of electronic digital computers were constructed from triodes, and in 1948 Claude Shannon wrote a paper laying out in detail just how mathematics should best represent and process information of all kinds. This paper helped guide the development of computers as reasoning machines. Today, computers constantly make decisions as well as calculations.

> SEE ALSO:
> ▶ Paradoxes of Zeno, Russell, and Gödel, page 166

A Minivac 601 was one of the first personal computers. It was developed by Claude Shannon in 1961.

Abstract Algebra

THE DEVELOPMENT OF ALGEBRA THROUGH THE AGES has been the story of increasing abstraction, and the single most important step forward was the introduction of a symbolic language by Diophantus and Viète. In the late 19th and early 20th centuries, this theme of abstraction was taken even further with the development of abstract algebra, and today abstract algebra is probably the most important branch of mathematics.

Its power lies in its abstraction. The more that algebra is unleashed from real, measurable phenomena, the better, because it can then be applied to a wider range of problems, including in economics, engineering, and quantum physics.

The equation $3^2+4^2=5^2$ would be useful for adding up the areas of two particular farmers' fields, but not for much else, and any new discoveries about this particular sum are unlikely to be very exciting. But the abstract $a^2+b^2=c^2$ is very useful, and anything new we could discover about it would be very interesting. Then $a^n+b^n=c^n$ takes us to a new level of abstraction, and the study of this equation led to the proof of Fermat's last theorem (page 98), and the development of powerful new areas of mathematics along the way.

All math, including algebra, arose from the need to measure real things. Abstract algebra unravels the structures of mathematics.

New words for new worlds

Finding out something new about $a^n+b^n=c^n$ would be a major breakthrough. But what if we could go further? What if we could find out something new about the concept of addition, or of raising to powers (exponentiation)? That's the kind of thing which would revolutionize the whole of mathematics, in a way that only a handful of mathematicians have ever achieved.

But there's a problem in even discussing this level of abstraction. In the last equation, a, b, c, and n all represent numbers. But if we're moving to a deeper level, that means we need a new language, with terms that do more than just represent numbers. This new language is made up partly of things called algebraic structures (see page 160).

The power of abstraction

All the new words involved make abstract algebra seem more complicated than it really is, especially because most introductions to the subject begin with long and precise definitions of the many concepts and structures involved. Often though, the easiest way to get to grips with the subject is through examples, which can be quite simple.

Emmy Noether is perhaps the most unsung of mathematicians. Her mathematics has driven much of the field of particle physics from the early 20th century to today.

Increasing abstraction makes algebra more and more useful for studying other subjects. Quadratic equations and other polynomials can be applied to engineering, economics, physics, and many other fields, but this is nothing compared to the power of abstract algebra. In areas like cosmology and string theory, it is the mathematical structures themselves that are explored, while Noether's theorem (see page 162) has led to new insights into the laws of nature. Group theory was used to predict both the Omega-minus particle, and the Higgs boson (see page 145).

Birth of a new mathematics

Some of the earliest ideas about group theory were produced by Cauchy, but the person who really showed the power of the subject was Galois. However, just as Newton discovered calculus because he needed a tool to explore the motions of objects, so Galois developed group theory mainly to explore the question as to whether quintic equations were solvable, and if not, why not (see page 140).

The person who is today recognized as the founder of abstract algebra is Emmy Noether, who was by far the greatest of all women mathematicians. In particular, she showed the great power of mathematical rings.

Outsider

Noether had a tough life, finding it very hard as a woman to get either a good education or a paid math job. Also, as a Jew in Germany in the 1930s, she was forced to flee to the USA as anti-Jewish attacks increased due to the growing power of the Nazis.

For the rest of her life, she was often treated with contempt, partly due to the fact that she was not interested in social convention—she wore comfortable clothes, said what she meant, and had no time for ladylike behavior. Yet, in fact, she seems to have been a very happy person, probably because only three things really mattered to her—family, mathematics, and her students, many of whom went on to become some of the greatest mathematicians of the 20th century.

Helping Einstein

Perhaps Noether's greatest breakthroughs happened when she took a break from pure math. In 1915, Albert Einstein was struggling to develop his theory of general relativity, and he asked her to help.

Einstein was trying to pin down the behavior of energy in time and space, and Noether was indeed able to help—and the experience got her thinking. The essence of most mathematicians

ALGEBRAIC STRUCTURES

Groups are a very powerful tool both for delving deeply into the underlying secrets of math, and for applying mathematical techniques to scientific problems. There are two pieces of information needed to set up a group:

1. The kind of numbers or shapes or objects that the group deals with.

2. The operation that is to be made on those numbers, shapes, or objects.

So, one group is formed of the integers, as operated upon by addition. It is usually referred to as "the integers under addition." It also has a symbol—$Z+$—and is an infinite group.

The operations involved here have to be those which just work on two numbers, shapes, or whatever. So, "Averaging" is not an operation that can be used to define a group, since averages can operate over many numbers at once. Two-number operations are called binary operations.

The numbers (or whatever) that are members of a group are referred to as a field. So we might talk about "the field of rational numbers." The rotations of a triangle make up a group called "the equilateral triangle under rotation." This is a finite group.

Groups can have subgroups: "the equilateral triangle under rotation" (which has three members) is a subgroup of the full symmetry group of an equilateral triangle, which includes reflections and has six members.

Abstract Algebra > 161

A graphical projection of the mathematical group called E8, a 248-dimensional object.

operation is only permitted to lead to other group members. So, for instance, "the integers under multiplication" is a group, because multiplying any integers together always gives another integer. But "the integers under division" is not a group, because, for instance, 1/2 = 0.5, and 0.5 is not an integer.

A ring is like a group, except that it includes two binary operations, such as addition and multiplication. (There's nothing particularly ring-like about rings, and it's not clear why they got this name.) Like quaternions, rings were initially invented in order to explore mathematics itself, but they are now essential to a lot of computer applications, including the system used to code information sent over the Internet.

Groups which describe the symmetries of triangles, or of any other polygon, are called dihedral groups. The symbol for the symmetry group of the equilateral triangle is D6, where the D is for dihedral, and the 6 is the number of members (or order).

If a group cannot be broken down into smaller groups, it is called a simple group. One of the essential points about a group is that its

The symmetries contained in a group can be added together to produce others.

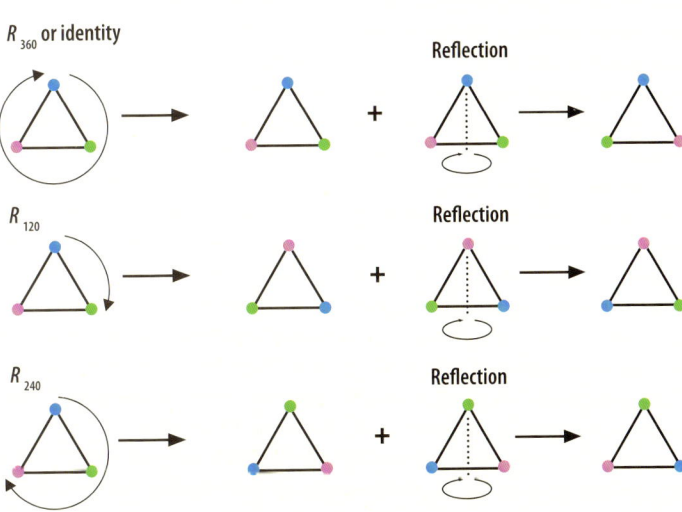

Even Noether's postcards to friends contained math.

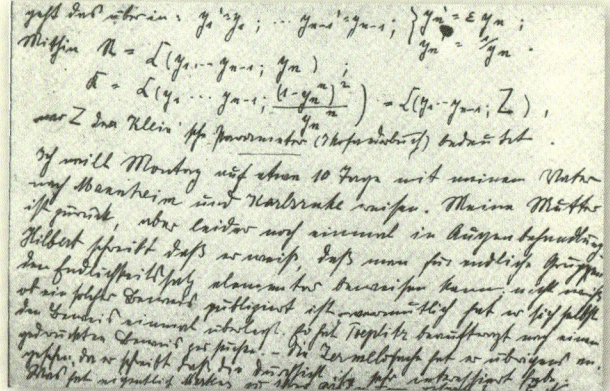

is the desire to generalize, and Noether was no exception. The theorem she developed is perhaps the most general and powerful in mathematical physics: it proves that every conservation law corresponds to a symmetry of nature, and vice-versa.

Conservation

In physics, some quantities are said to be conserved. That is to say, they can never be destroyed, though they may change form. Energy is one example. Nuclear energy in the Sun changes to light and heat energy. The light is absorbed by plants, which store it as chemical energy. We eat the plants and digest them, freeing the energy and using it to move, that is, it becomes kinetic energy. This motion heats up our bodies and our surroundings as that kinetic energy turns into heat energy. But the amount of heat energy at the end of the process is precisely the same as the amount of nuclear energy at the start. Momentum is also conserved. If one object strikes another, the first slows (losing momentum) and the second starts to move (gaining momentum).

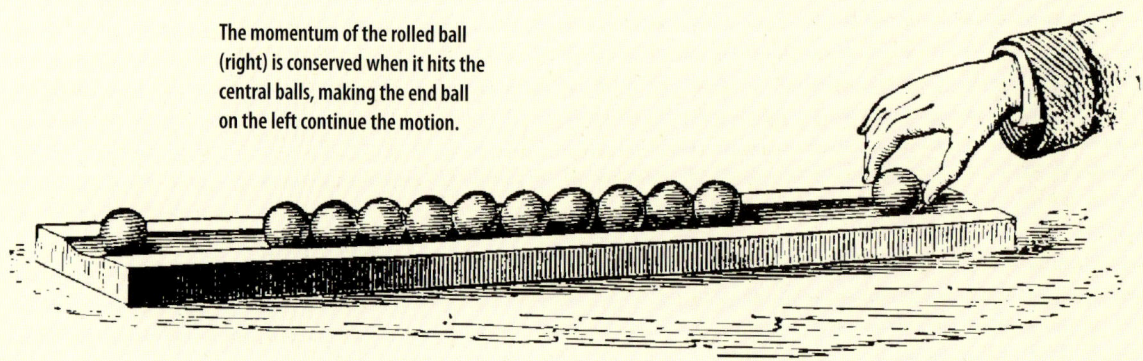

The momentum of the rolled ball (right) is conserved when it hits the central balls, making the end ball on the left continue the motion.

How it works

The momentum of an object is its tendency to remain in motion. If you catch a ball thrown at you, you can feel this effect: It takes some effort to overcome the ball's tendency to keep going. The heavier the ball is, the more difficult it is to stop, which tells us that momentum is proportional to mass (momentum$\propto m$). Momentum depends on velocity, too. The faster the ball is moving, the harder it is to stop (momentum$\propto v$). These are the only two things that momentum depends on, so we can conclude that momentum = mv.

Usually, it is easy to see that momentum is conserved: When you catch a ball, the momentum is transferred to your hand. Some of it moves your whole hand back a little, while some of it is transferred to molecules in your hand, speeding them up—which you can feel as a slight warmth if the ball hits you hard enough. (Without momentum, sports like baseball would be no fun at all.)

But there are cases where it seems that momentum is not conserved. If you drop a ball, it will get faster as it falls, rapidly increasing its momentum. We can explain this by saying that the ball is actually part of a larger system which includes the whole Earth, and the total ball + Earth momentum remains the same. This is the same idea as with a ball that is caught—if we could only see the ball, not the catcher, it would look as if the ball's momentum just vanished. However, who is to say which objects are part of which systems? The secret is symmetry. Bearing in mind that a symmetry is a change that leaves a system ending up looking the same, let's look at the ball again, and make a change to it. Let's change its position. Imagine the ball, far away in outer space. It hangs motionless. We don't want to move the ball as that would give it extra momentum, so let's place another identical ball a few miles away. Both will hang motionless, their momentums identical. So, this is an example of a symmetry, one related to position. Now, take another identical ball but place this one about 100,000 miles above the surface of the Earth. This time, it will begin to move as soon as we release it, slowly drifting toward the Earth. If we placed another identical ball just 1,000 miles above the Earth, it would also start to move, much faster this time. So, there is no symmetry here.

The reason is that the space near a massive object is different to space far away from it. Close to a massive object, space is distorted and asymmetric, and momentum is not conserved within any part of that space. Only if the whole region is considered, including the massive object itself and space for many millions of miles around, does the conservation of momentum apply.

BROKEN SYMMETRIES

Every magnet has an opposite pole at each end, so magnets are not symmetrical objects. If you have two of them lined up so that their north poles face one another, they will try to push each other apart. But, if you rotate one of them, so that a south pole now faces the north pole of the other, there will be an attraction between them. So, that change leaves you with a different situation to the one you started with, and therefore is not an example of a symmetry.

However, if you heat the magnets up sufficiently (past a temperature known as the Curie Point), the magnets will lose their magnetism. Now, they have no poles and you can rotate them without changing their relationship. So, now there is a symmetry where none existed before. But the magnetism is not lost forever. If you cool the magnets, they can be re-magnetized. This cooling and remagnetizing is an example of symmetry breaking.

In the early Universe, very soon after the Big Bang, temperatures were enormous and it is believed that magnetism did not apply. More strangely, neither did gravity or the other basic forces of nature (electromagnetism and the strong and weak forces). At high enough temperatures, there was no way to tell the difference between them. This is really amazing, considering that gravity and magnetism are so different. Gravity affects everything the same way in all directions. Magnetism only noticeably affects metals and a few other things, and can attract as well as repel. It also pairs up with electricity and delivers us motors and dynamos, wind farms and radio waves, none of which happens with gravity.

Mathematics helps us study these very early phases of the Universe, when the temperatures were so high that we can't control them in laboratories.

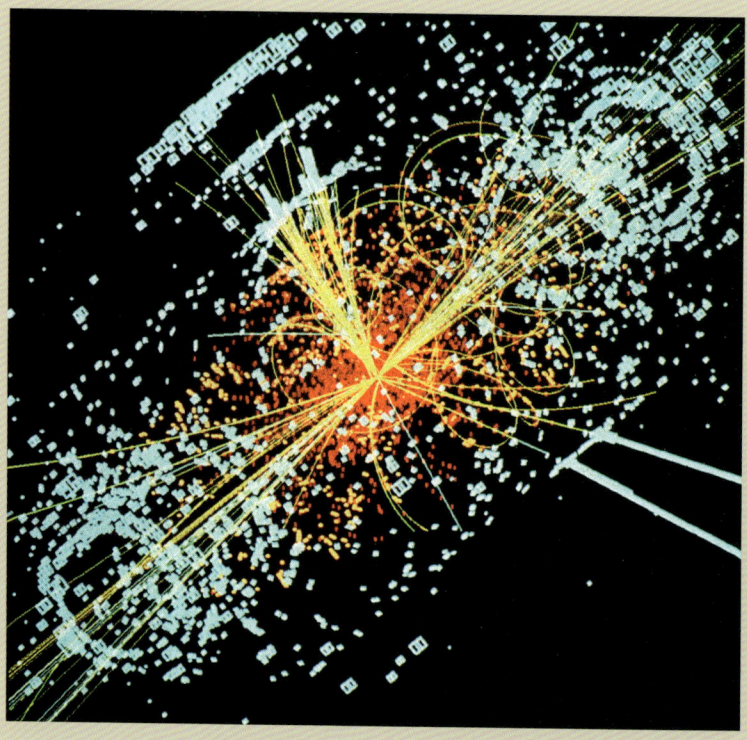

An early phase of the Universe is described as the quark-gluon plasma, which is recreated and imaged inside particle accelerators.

Conservation laws, as they are called, are about the most fundamental there are. Discovering them was a real triumph for physics. But we can ask the same question we did earlier about math: Can we go further? Can we explain *why* some things are conserved, while others (like light, water, or sound) are not?

Thanks to Noether, we can. Her theorem is not only very useful in itself and is used routinely in many areas of science, from engineering to cosmology, but also her approach, together with those of other group-theorists, led to a powerful

Einstein's theory of relativity describes how space and time are warped by energy. His theory is based on the speed of light being fixed ("absolute") so space and time must be changeable ("relative").

new way of looking at basic areas of physics. In particular, physicists who are investigating the beginning of the Universe or the fundamental nature of matter, make a lot of use of the concept of broken symmetries.

SEE ALSO:
▶ Finding the Maximum, page 86
▶ Groups, page 140

Paradoxes of Zeno, Russell, and Gödel

IN THE HISTORY OF MATHEMATICS, there have been three great crises, and mathematicians responded in three very different ways.

The first and most important of all these crises was the discovery of irrational numbers in about 530 BCE (see page 30). This changed the whole course of mathematics, moving the focus away from algebra to geometry, where the subject remained for generations. Irrational numbers were only accepted as valid parts of mathematics centuries after they were discovered. The second crisis struck at the foundations of calculus. The concept of infinitesimals as things which were almost but not quite zero had been uncomfortable for many mathematicians from the very beginning, but most found calculus so useful they were willing to accept them. But as math evolved and calculus became ever more important, nagging doubts developed into criticism. It was a philosopher and bishop, George Berkeley, who perhaps did most to sum up these difficulties and make mathematicians take them seriously, largely by making fun of infinitesimals, calling them "the ghosts of departed quantities." Infinitesimals can be regarded as the result of making a number grow ever-smaller. But what happens in the end?

Zeno of Elea baffles his peers with his paradoxical mathematics.

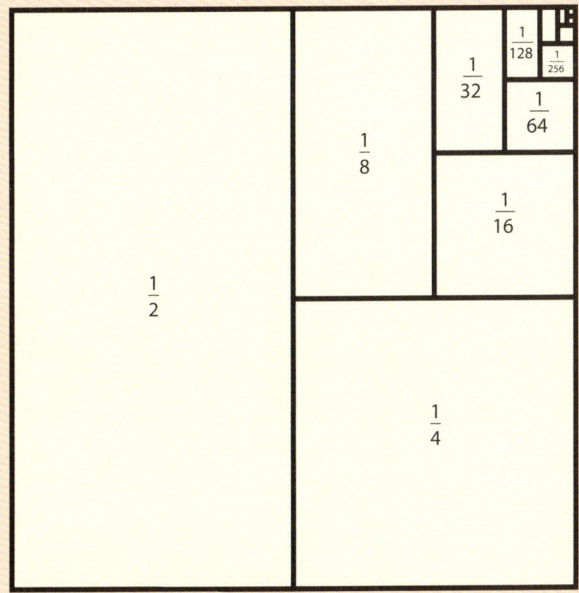

This square can be divided into infinitely smaller spaces—potentially—but in reality that is impossible.

Achilles and the tortoise

It was an ancient Greek called Zeno of Elea who first pointed out why this question was important, when he challenged people to explain how even a champion runner like Achilles could ever overtake a slow creature like a tortoise, if the tortoise has a slight lead. Let's say it takes Achilles half a second to get to the point where the tortoise was when he set off. By this time, the tortoise will have moved on a short distance. Perhaps it takes Achilles a quarter second to cross this distance. But by then, the tortoise will have moved again. Achilles will need another eighth of a second. There seems no end to this process, which implies that even Achilles will never reach the tortoise.

One answer that appealed to the Greeks was the theory that everything is made of atoms, tiny things too small to see that cannot be divided ("*atmos*" means "uncuttable"). If that was true, then Achilles would reach a point where he could no longer half the distance any more, and—presumably—could just jump over it.

Infinite difficulty

For mathematicians there was a very similar problem with numbers. Take Kepler's barrel problem again (page 86). There seems no reason, except boredom or lack of time, why one shouldn't go on with the process forever, getting closer and closer to the result. In that case, the cask could be regarded as being made of an

Democritus took life (and death) less seriously than his contemporary philosopher Heraclitus.

infinite number of very thin slices. But now we are back with a Zeno-like problem again. If there are an infinite number of slices, how big are they? If they are a trillionth of a trillionth of a cubic inch each, then an infinite number of them will add up to infinity (infinity times any number = infinity). On the other hand, if each has a zero volume, then even an infinite number of them will add up to zero (0 + 0 + ... = 0). So if there must be an infinite number of them (because otherwise there would have to be some reason for them to stop), and they can't be finite in size, and they can't be zero either ... then what?

$$\lim_{x \to \infty} \frac{1}{x} = 0$$

$$\lim_{x \to 0^+} \frac{1}{x} = \infty$$

Setting limits

Even if we could somehow get Achilles past the tortoise by talking about indivisible atoms, that is no help with numbers. The question came down to this: could infinitesimals be properly defined, so that their behavior could be understood? Or, alternatively, if in fact they could not be defined, was it possible to base calculus on something different? This problem occupied mathematicians for many years, but was largely resolved by Augustin-Louis Cauchy. First, he defined a limit, like this: "When the successive values attributed to a variable approach indefinitely a fixed value so as to end by differing from it by as little as one wishes, this last is called the limit of all the others." Then, he defined an infinitesimal as "A variable which decreases indefinitely in such a way so as to converge toward the limit zero."

In this way Cauchy avoided the unanswerable question: "If you shrink something until you

The mathematical notation used to describe nothing from something and everything from nothing.

can go no further, what do you end up with?" He simply said, "Just shrink it as much as you need." Or, one could say he replaced the concept of "infinite smallness" with "whatever amount of smallness does the job."

Another crisis

The third crisis took place at the beginning of the 1930s. By that time, mathematicians had become increasingly aware that, in fact, many of the proofs of earlier mathematicians were not as certain as they could be, and there was a feeling that mathematics should rest on absolutely firm foundations. The two mathematicians in particular who were keen on this idea were David Hilbert (who, in 1900, laid out a series of

23 questions that he thought should define the problems that mathematicians should tackle in the new century) and Bertrand Russell, a philosopher and mathematician who wanted to show that all math was logic in the end. Together with a colleague, Alfred Whitehead, Russell produced an enormous three-volume work that was intended to do for math what Isaac Newton's *Philosophiæ Naturalis Principia Mathematica* (Mathematical Principles of Natural Philosophy) had done for physics. It even had a similar Latin title, *Principia Mathematica*. But there was a problem.

Paradoxes

As a philosopher, Russell was very aware of the existence of paradoxes. Statements which, although they seem sensible at first glance, actually make no sense. Here's an example:

**The next sentence is true.
The last sentence is false.**

The question is, which, if either, of these sentences are true? If the first one is true, then it says the next one is true. So, the next one must be true. But the next one says that the first one is false. So, the first one must be false. So, the first sentence should read "The next sentence is false." But that would mean that the second sentence should really read "The last sentence is true"… and we're back where we started, going round and round in a loop of logic. There are many examples like this, even the simple sentence

This sentence is false.

is one of them, as we can see if we try to work out whether it is true or not. If it's true, then it's false, because it says it is. But that means it's true. So it's false …

Making sense of nonsense

Now, it may well be that all this shows is that language is a funny thing, not very scientific, and you can make it turn out nonsense. But then, who says language has to be scientific? Poetry, for example, uses language in a very non-scientific way and is all the better for it. But the problem for Russell and his colleagues was that this kind of paradox seemed to apply to the work they

Bertrand Russell was a powerful force in philosophy as well as math, and was a global voice advocating pacifism.

were doing too (see box, opposite). It was for this reason more than any other that Russell and Whitehead decided to write their *Principia*, to ensure that mathematics could avoid the paradoxes that so concerned him.

Total destruction

All was well until 1931, when a 25-year-old German mathematician called Kurt Gödel resolved the issue of paradoxes in mathematics—but not in a way that pleased everyone. What Gödel's theorem did was to analyze "formal systems" in general—and that included mathematics—to see whether there were paradoxes in them, and if so, what that meant. He proved that there *are* paradoxes in mathematics, and that has a devastating implication.

Above: Russell's *Principia* did not prove $1 + 1 = 2$ till page 379.

Right: Kurt Gödel, center, receives an award from Albert Einstein.

PARADOXES IN MATHEMATICS

Mathematics makes a lot of use of sets. The word "set" doesn't have a precise definition, and it just means a collection of something—so, the integers are a set, and so are the planets, or the books in a library. The problem that worried Russell concerned sets.

1. Most sets are not members of themselves. The set of ships is not a ship and the set of numbers is not a number. 2. But there are some sets which are members of themselves. The set of "mathematical ideas" is a member of itself, the set of "things with the letter 's' at the beginning" is a member of itself, and the set of sets is a member of itself, too. 3. We can make sets out of anything at all, so let's define a new set, which will be "the set of sets which are not members of themselves." So, this will include the set of ships and the set of numbers, but it will not include the set of mathematical ideas, the set of sets, or the set of things beginning with "s." 4. Is this new set a member of itself? 5. If it *is* a member of itself then it cannot be part of "the set of sets which are *not* members of themselves." Therefore it is *not* a member. 6. If it is *not* a member of itself, then it must be part of "the set of sets which are not members of themselves." Therefore it *is* a member.

Here is a slightly easier example. In a small village, there is a barber called Joe. Most men in the village shave themselves, but some prefer to go to Joe, and he shaves them. In fact, he shaves all the men in the village who do not shave themselves. One day,

Joe gets worried by a rumor that another barber is coming to live in the village. Joe goes to the village council and convinces it to pass a law that will only allow Joe to offer shaves. There is some concern about this though—the council is worried that Joe might try to make a bit of extra money by using the law to insist on shaving men who have already shaved themselves. Or, what if Joe should put his prices up a lot? His customers couldn't go to another barber, now that Joe is being officially made the only one. So, although the council does pass the law, it is worded very carefully, like this: "Joe the barber *must* shave every man in the village who does not shave himself (these men will be referred to as 'self-shavers'). But Joe the barber *must not* shave any self-shavers. Anyone breaking this law will be fined."

Joe is pleased, but, the next morning, he goes to his bathroom as usual, picks up his razor, and … stops. If he has a shave then obviously he is a "self-shaver." But the law says Joe the barber must not shave any self-shavers. Joe puts his razor down again. Maybe he will have to grow a beard. Then he definitely will not be a "self-shaver." But then he remembers the law. Joe must shave every man who is not a self-shaver. He picks up the razor again …

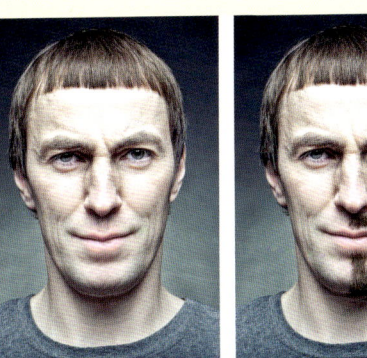

Gödel's ideas led to powerful calculators such as the Bombe, used to crack the German codes during World War II.

Going back to the example "This sentence is false," we can see that it refers to itself (is "self-referential"). Self-reference is what Russell had worked so hard to avoid in terms of sets. What Gödel did was to explore a self-referential statement like this:

Statement S: "Statement S cannot be proved."

Now, it certainly seems an odd statement, but no mathematician need worry about that because it doesn't have anything to do with math. But what if it could be written as a mathematical statement? Gödel's next step was to turn it into one. The details are complicated, but the basic idea was to convert every part of the statement to numbers. This is a bit like taking the sum "2 + 2 = 4," and assigning numbers (called Gödel numbers) for the + and = signs, too. So, we might end up with 2662994.

Russell and other mathematicians had already invented symbols to stand for words like "proved," "statement," and "cannot," so Gödel's task wasn't as hard as it might seem. In fact, it was key to his proof that he used the same rules as Russell had when he wrote *Principia Mathematica*. So, the enormously long numbers that Gödel ended up with had been formulated in strict accordance with the rules of mathematics. It was just as "mathematical" as "2 + 2 = 4."

This didn't mean it was true, of course: "2 + 2 = 5" is also a mathematical statement, it just happens to be a false one.

True and false

Let's imagine the Gödel number for Statement S: "Statement S cannot be proved," is 123456789. Is 123456789 true? Let's assume it is. If so, then, as it says itself, it cannot be proved. So, we have an example of a true mathematical statement that cannot be proved. This is itself a major problem, since Gödel was able to show that the existence of just one such statement meant that there could be countless others. Many mathematicians trying to prove new things in mathematics might be wasting their time—maybe Fermat's last theorem was true, but unproveable.

But perhaps all was well. Maybe 123456789 is false. But look what it says: "Statement S cannot be proved." If that is a false statement, then that means Statement S can be proved. In which case, we have found a way to prove something false. Again, Gödel managed to show that the existence of this one example of a false-but-provable statement meant that there were countless others. Which meant this was an even more awful conclusion! Maybe some of the tried and trusted theorems and formulas that mathematicians relied on, and which had been proved, were actually false.

Gödel didn't stop there. He proved that any mathematical system would inevitably allow the construction of statements which, although easy enough to understand, would be neither true nor false nor meaningless. What's more, he managed to prove that these statements could not be identified—they would not necessarily be self-referential statements like Statement S. There was (and is!) no escape. Either mathematicians must accept that there are statements which are neither true nor false (in which case mathematics is said to be incomplete), or it must say that they are both true and false (in which case, mathematics is said to be inconsistent). This was devastating. Russell's plan was in ruins—not only had he failed to prove that mathematics was free of paradoxes, it now turned out that it was full of holes, too. No one has ever found an escape from Gödel's discoveries. In fact, several promising theorems have turned out to be unprovable, and mathematics is not, and can never be, the perfect and complete structure that most earlier mathematicians believed it should be.

> SEE ALSO:
> ▶ The Mathematics of Thought, page 152

Alan Turing devised a system to investigate Gödel's incompleteness theorem—and ended up inventing the general-purpose computer.

CLAY MATHEMATICS INSTITUTE

7 Millennium Problems

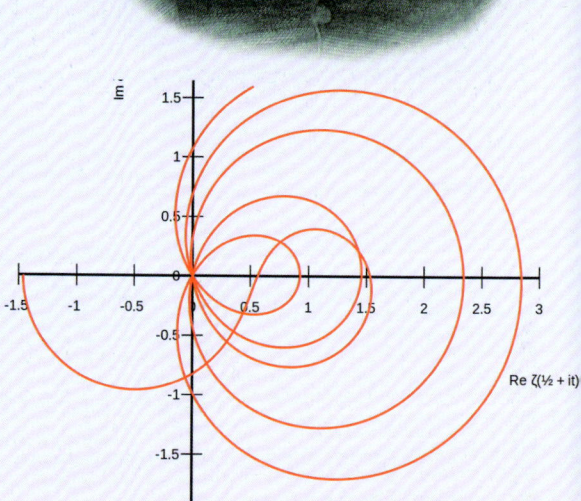

The Riemann Hypothesis is named for Bernhard Riemann, who proposed it in 1859.

DESPITE THE IMPACT OF GÖDEL'S THEOREM ON THE HOPES AND DREAMS OF MATHEMATICIANS, mathematics since then has leapt forward in many areas, thanks to a combination of easier access to education, better ways to share information, and the application of mathematics to fields like biology, medicine, and criminology.

In 2000, the international Clay Mathematics Institute selected the following seven Millennium Problems, offering a prize of $1 million to whoever can solve one. They illustrate some of the main areas of interest for this centuries' mathematicians. Only one has so far been solved.

Riemann's zeta function is a tool for investigating prime numbers.

Yang–Mills and Mass Gap
Area: Group theory applied to physics.
Investigates: The masses of sub-atomic particles.

Riemann Hypothesis
Area: Prime numbers.
Investigates: Patterns in the infinite sequence of prime numbers.

P vs NP Problem
Area: Computing.
Investigates: If it is easy to check the solutions of a problem, is it also easy to solve the problem?

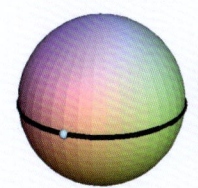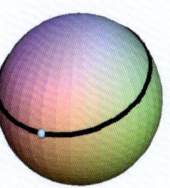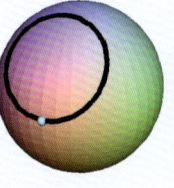

The Poincaré Conjecture concerned how shapes form on shapes with more than three spatial dimensions.

Navier–Stokes Equations (NSEs)
Area: Differential equations.
Investigates: Whether the NSEs always work.

Hodge Conjecture
Area: Algebraic equations and topology.
Investigates: Assembling shapes from building-blocks with different numbers of dimensions.

Poincaré Conjecture (SOLVED in 2002)
Area: Geometry.
Investigates: Three-dimensional surfaces.

Birch and Swinnerton-Dyer Conjecture
Area: Number theory.
Investigates: The solvability of equations.

But investigations into the Millennium Problems are only some of a vast range of research projects in mathematics. One of the most wide-ranging, fundamental, and challenging of these projects is the Langlands Program, which seeks to relate together areas of mathematics that seem very different, so that solutions developed for one area can be applied to problems in another.

Though being a professional mathematician is one of the most rewarding jobs there is, it is not only professionals who make breakthroughs. For instance, while many people have studied the ways in which two-dimensional shapes can link up to cover a surface, four of these ways were found by Marjorie Rice, an amateur mathematician, in the 1970s.

Today, mathematics is more accessible than ever before, thanks largely to the Internet. YouTube has a wealth of excellent mathematical videos, social networks support international communities of mathematicians, and for professional mathematicians, the online platform arXiv is by far the most popular way to present and discuss the latest research.

Mathematics has never been easier. Get going!

Even supercomputers take almost forever to solve NP (non-polynomial) math problems. Time to think differently.

Calculus in Depth

FOLLOWING OUR LOOK AT THE HISTORY OF CALCULUS on page 110, here is a further tour of the more advanced techniques that give this mathematics such power.

Second differentials
Since a differential is a rate of change, it often makes sense to differentiate more than once. Given a formula for the position of a moving object, we can find the rate of change of that position (that is, its velocity) by differentiating the formula. If we differentiate again, we will get a formula for the rate of change of velocity, which is acceleration.

So, if an accelerating racing car's distance from its starting point in feet is given by $4t^2$ (where t is time in seconds), we can differentiate once to find the formula for its velocity, and twice to find its acceleration formula:

Position: $x = 4t^2$;
Velocity: $v = \frac{dx}{dt} = 8t$;
Acceleration: $a = \frac{dv}{dt} = 8$

Instead of doing this in two stages, we can just find the "second differential" of the position, which is written like this:

$\frac{d^2y}{dx^2}$ and is given by
$\frac{d^2y}{dx^2} = (n-1)ax^{(n-2)}$

Partial differentiation
What if we want to find the slope at a selected point on a three-dimensional surface, rather than a two-dimensional curve? For three-dimensional spaces, we need to introduce a third variable, z. Let's plot the surface represented by $z = -0.5x^2 + y^2$.

Also, let's say we want to find the slope at the point with coordinates (42, 46, 1234). As usual, we need to be precise in our question: There are many slopes that pass through this point, all with different values, and we can imagine the surface being made up of many curves, very close together. We can think of those curves as all being parallel to the y-axis. The first of those curves is the one that runs along the right-hand edge of the surface.

But we can equally well think of a set of close-together curves, all running parallel to the x-axis. The point we are interested in lies on (at least) two curves, one which is parallel to the x-axis and one which is parallel to the y-axis. And each of those curves has a different slope at that point. How do we find those slopes?

The simplest way to approach this is to avoid differentiating the function all at once. Instead, we first differentiate it with respect to x, while ignoring the variable y (or rather, keeping y constant), and then differentiate it with respect to y, this time keeping x constant. This is called partial differentiation, and we write the two differentials like this:
$\frac{\partial z}{\partial y}$ and $\frac{\partial z}{\partial x}$

Now we can use the standard $nax^{(n-1)}$ expression
$z = -0.5x^2 + y^2$,
so $\frac{\partial z}{\partial x} = -(2 \times 0.5)x = -x$

(Note that the y^2 disappears when it is differentiated, just as any lone constant would). And now, to differentiate with respect to y, we use effectively the same expression, $nay^{(n-1)}$:

$\frac{\partial z}{\partial y} = 2y$

So, at the point (42, 46, 1234), the slope in the x-direction is $-x$, which is -42, and the slope in the y-direction is $2y$, which is $2 \times 46 = 92$.

Selecting the integration method
Integrating a more complicated function is not a matter of plugging values into a standard formula; there are a number of different methods that might work, but there are no reliable rules to decide which is best. In other words, integration is a matter of experience, skill, and luck.

Integration by trigonometric identity
Integrating (or differentiating) simple trigonometric functions can be done

just by looking up the standard values, which are:

Function	Integral	Differential
$\sin\theta$	$-\cos\theta+c$	$\cos\theta$
$\cos\theta$	$\sin\theta+c$	$-\sin\theta$
$\tan\theta$	$-\ln(\cos\theta)+c$	$\sec^2\theta$

If we wish to integrate a function where there is an angle multiplier, like $\sin 6\theta$, the answer must be divided by that multiplier, so $\int \sin(6\theta)d\theta = -1/6 \cos(6\theta)+c$. Armed with knowledge of these, and of trigonometric identities such as $\sin(2x)=2\sin x \cos x$, we can sometimes simplify an integration problem. For instance, to find

$$\int_0^{\pi/4} 2\sin x \cos x \, dx$$

We can use the identity $\sin(2x) = 2\sin x \cos x$

So, we just need to find

$$\int_0^{\pi/4} \sin(2x) dx$$

Which is

$[-1/2\cos(2x)+c]_0^{(\pi/4)}$

And that is

$(-\cos(2\times\frac{\pi}{4})+c - -\cos(2\times 0) - c)$

$= 0 - 1 = -1$

Integration by trigonometric substitution

This approach also needs knowledge of the integrals of trigonometric functions, and of trigonometric identities.

For instance, in

$$\int \frac{1}{\sqrt{9-x^2}} dx$$

The $9 - x^2$ expression in the square root has a similar form ("constant minus variable squared") to the trigonometric identity $1- \sin^2\theta = \cos^2\theta$

But we need to change the 9 into a 1. We can do this by substituting the x^2 with $9\sin^2\theta$.

So,

$$\int \frac{1}{\sqrt{9-x^2}} dx \text{ becomes}$$

$$\int \frac{1}{\sqrt{9-9\sin^2\theta}} dx$$

However, before we can go further, we need to change all of the x terms in the expression into θ terms, which means we need to find an expression for dx. Let's look at the substitution we have already chosen: x^2 has become $9\sin^2\theta$. So, x must be $3\sin\theta$. We again need knowledge here of the differentials of trigonometric functions, which will tell us that the differential of $\sin\theta$ is $\cos\theta$ (and the differential of $3\sin\theta$ is $3\cos\theta$). That is:

$\frac{dx}{d\theta} = 3\cos\theta$

So, we can rearrange this to get what we wanted: The expression in terms of θ that we need to replace our dx with, in order to complete our substitution. That expression is: $dx = 3\cos\theta d\theta$.

Now, we can go back to our

$$\int \frac{1}{\sqrt{9-9\sin^2\theta}} dx$$

and replace the dx with $3\cos\theta d\theta$, to get

$$\int \frac{3\cos\theta}{\sqrt{9-9\sin^2\theta}} d\theta$$

Now that we've substituted for all the x terms, we can start to simplify the expression. First, we can take the 9s out of the square root sign, like this:

$$\int \frac{3\cos\theta}{3\sqrt{1-\sin^2\theta}} d\theta$$

(the 9s in the root have turned into a 3 outside the root)

And now we can cancel the 3s to give

$$\int \frac{\cos\theta}{\sqrt{1-\sin^2\theta}} d\theta$$

Now we have isolated the expression $1- \sin^2\theta$, so we can use our $1-\sin^2\theta =\cos^2\theta$ trigonometric identity:

$=\int \frac{\cos\theta}{\sqrt{\cos^2\theta}} d\theta \ =\int \frac{\cos\theta}{\cos\theta} d\theta \ =\int d\theta = \theta+c$

And we finish the integration by converting back this final expression to an expression in terms of x, as follows.

Our substitution involved replacing x by $3\sin\theta$. So, $\sin\theta=x/3$. Hence, $\theta+c$, which is the solution we've found to our integral, must equal $\arcsin(x/3)+c$. And that's the answer we were looking for.

Integration by u substitution

To integrate a function like this one:
$y = \int(x+3)^6 dx$

we could multiply out the brackets, and then integrate the terms that result. But there is a much faster way; we temporarily replace the expression inside the brackets with any variable symbol (u is usually used); then integrate the function with respect to u:

i) Define $u = x+3$

ii) Replace the $x+3$ in the function with u:
$y = \int (u)^6 dx$

iii) Work out what the dx is. We do this by differentiating the expression for u:

$u = x+3$, so $\frac{du}{dx} = 1$

We can rearrange this, giving $dx = du$

iv) Replace the dx in the u-version of the function with du
$y = \int (u)^6 du$

v) integrate

$\int y\, du = \frac{1}{7} u^7 + c$

vi) change the u back into the $x+3$ again, to give the final answer

$\int y\, dx = \frac{1}{7}(x+3)^7 + c$

Integration by parts

If we want to integrate a product, integration by parts often works. It is based on this identity
$\int u\, dv = uv - \int v\, du$
For instance, if we want
$\int x \cos x\, dx$

First we change the x into u and the $\cos x\, dx$ into dv, so now we have $\int u\, dv$.

Now we need to find du and v

$u = x$, so $\frac{du}{dx} = 1$, so $du = dx$

since $dv = \cos x\, dx$, we can find v by integrating this expression

$v = \int \cos x\, dx = \sin x + c$

Now that we have expressions for u, v, du and dv, we can put them into our integral

$\int u\, dv = uv - \int v\, du$
$\int u\, dv = x(\sin x + c) - \int \sin x + c\, dx$
$= x \sin x + \cos x + k$

Integration by partial fractions

If the expression to be integrated is in the form of a fraction, it may be possible to simplify the problem by breaking it into separate, "partial" fractions. But, any expression which is a fraction with the variable in the denominator (the part underneath the line) yields a natural logarithm when it is integrated. Luckily, this is fairly simple:
$\int \frac{1}{x}\, dx = \ln(x)$

So, if we have this expression to integrate:

$\frac{3x + 11}{x^2 - x - 6}$

It can be represented by these partial fractions, which are easier to integrate
$\frac{4}{x-3} - \frac{1}{x+2}$

So, we just integrate those two terms

$\int \frac{4}{x-3} - \frac{1}{x+2}\, dx$

Which gives
$4\ln(x-3) - \ln(x+2) + c$

Numerical integration: Rectangles

The above six integration methods are referred to as "analytic." Because there are so many of them, and skill is needed to choose the right one, it is very challenging to program a computer to carry out analytic integration. There are also some functions that can't be integrated by any of these methods, including most of those that turn up in physics, biology, and engineering.

Fortunately, there is an approach which works for almost any function: Numerical integration. Numerical methods are simpler to describe than most analytic ones, and little skill is needed to use them. This means that it is relatively easy to write a step-by-step method (algorithm) that a computer can follow. Numerical methods involve a great many calculations which are extremely tedious, but fortunately tasks like that are exactly what computers were invented to carry out for us. So, while human mathematicians almost always use analytic methods if they possibly can, computers almost always use numerical ones.

While analytic methods result in an expression (an indefinite integral), which, if required, can be converted to numbers (a definite integral), numerical methods yield numbers in the first place. Hence, all numerical integrals are definite integrals.

Numerical integration is easiest to visualize when the function that needs integrating has been displayed as a graph. In that case, finding the integral means finding the area under the curve.

The approach is always the same: Cover the area of interest with shapes that you know how to find the areas of, find those areas, and add them together. Although you could use a whole menagerie of shapes, it's much easier to chose just one sort of shape and cover the area with variations of that.

Various shapes are used, but the simplest is the rectangle.

For example, to integrate the area under the curve defined by $f(x) = 0.01x^2 + 0.1x + 100$ between $x = 0$ and $x = 100$:

First, we decide how many rectangles we will divide the area into. The more we use, the closer the fit will be, but the more calculations we will have to do. Since it's we humans who are doing this one, rather than tireless computers, let's use just five strips (we'll check the closeness of the fit this gives us later).

The best fit will be if the curve passes through the midpoint of the top of each rectangle. These are the x-positions of the start, midpoint, and end of each strip-top.

Strip	Left	Midpoint	Right
1	0	10	20
2	20	30	40
3	40	50	60
4	60	70	80
5	80	90	100

Next, we need to know the heights of the rectangles. We get these from the equation for the line $(f(x) = 0.01x^2 + 0.1x + 100)$, plugging

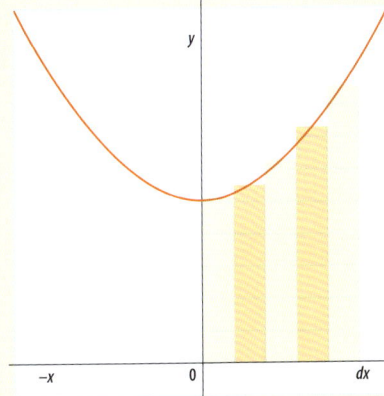

In numerical integration, an area under the curve is divided into strips. Their areas are added up. The narrower the strips are, the more accurate the answer, but the longer the calculation becomes.

into it the midpoint values from our table.

x	f(x)
10	102
30	112
50	130
70	156
90	190

And now we simply work out the five areas and sum them. The width of each strip is 20, and their heights are the midpoint values calculated above, so the total is $(20 \times 102) + (20 \times 112) + (20 \times 130) + (20 \times 156) + (20 \times 190)$

We can simplify this:
$20 \times (102 + 112 + 130 + 156 + 190)$

Which is 13,800 square units.

In this example, we can check our answer by integrating the original function, $f(x) = 0.01x^2 + 0.1x + 100$

$$\int_0^{x=100} df(x) = \left[\frac{0.01x^3}{3} + \frac{0.1x^2}{2} + 100x + c\right]_0^{100}$$

$$= \left(\frac{0.01 \times 100^3}{3} + \frac{0.1 \times 100^2}{2} + 100 \times 100 + c\right)$$

$$- \left(\frac{0.01 \times 0^3}{3} + \frac{0.1 \times 0^2}{2} + 100 \times 0 + c\right) \approx 13,833$$

So, this is about 33 (about 1%) different from the value from our numerical integration.

To write a general formula for numerical integration by rectangles, we just need an expression to describe the area of any strip: If the area of interest starts at $x = a$ and ends at $x = b$, and if we have n strips, then each strip has the width $(b-a)/n$. Let's call the edges of the rectangles $x_0, x_1 \ldots x_5$. The horizontal midpoints of the rectangles are then at $(x_0+x_1)/2$, $(x_1+x_2)/2, \ldots (x_4+x_5)/2$. The heights of the midpoints are defined by the value of the original function at these points, that is $f(x_0+x_1)/2, f(x_1+x_2)/2, \ldots f(x_4+x_5)/2$. If we want to make this more general, so that it works for any number (n) of strips, we just keep going until we reach the final value, which is $(x_{n-1}+x_n)/2$.

The symbol for the sum of a series is Σ (a capital Greek S, for sum), and to sum a number n of items from 1 to k, it takes the form $\Sigma_{k=1}^n$

So, putting that all together, the expression for the total area is

$$\Sigma_{k=1}^n \frac{b-a}{n} \left[f(x_0+x_1)/2 + \ldots + f(x_{n-1}+x_n)/2\right]$$

Glossary

Coefficient
A number which multiplies a variable. In the equation $6 = 3x$, 3 is the coefficient.

Complex number
A number of the form $a + bi$, where a and b are real numbers, and i is $\sqrt{-1}$.

Conjecture
A mathematical statement which has been suggested but not proved. Once proved, it becomes a theorem.

Constant
A fixed value in an equation. In the equation for a straight line, $y = mx + c$, c is a constant, m a coefficient and x and y are variables.

Coordinate
A group of numbers which defines a position.

Cubic equation
A polynomial of degree 3, such as $x^3 - 4x^2 + 1 = 0$.

Degree (of a polynomial)
The largest exponent of a polynomial. The polynomial $x^7 - 3x^4 = 0$ is of degree 7.

Differential equation
An equation that contains both a function and its differential, such as
$$\frac{dN}{dt} = rN\left(1 - \frac{N}{k}\right)$$
Many laws of physics, biology, chemistry, and economics can be written as differential equations; this example gives the rate of increase of the number (N) of rabbits or other animals over time (t) in terms of their breeding rate (r) and the amount of lettuce available (k).

Differentiation
A method of finding the rate at which something changes. For instance, the acceleration of a car can be found by differentiating a series of values of its velocity.

Exponent
The number or expression showing the power to which another number or expression is to be raised. 2^{2n} means 2 should be multiplied by itself $2n$ times.

Function
An equation into which you can put an input to get an output, like $y = x^2$. x is the input here, and y is the output, so if you input $x = 2$, you get the output $y = 4$.

Integer
A number in the series … -3, -2, -1, 0, 1, 2, 3 …

Integration
The opposite of differentiation, which can be used to find the areas under curves or the volumes of solid objects.

Irrational number
A number which cannot be expressed as a ratio (fraction).

Natural number
A counting number: 1, 2, 3, 4 ...

Polynomial
A mathematical expression that contains only variables, coefficients, exponents, constants, and the operations of addition, subtraction, and multiplication (though not necessarily all at once). The word "polynomial" means "many numbers."

Prime number
A number which can be divided only by itself and by 1 to leave no remainder.

Quadratic equation
A polynomial of degree 2.

Quartic equation
A polynomial of degree 4.

Quintic equation
A polynomial of degree 5.

Rational number
A number which can be expressed as a ratio or fraction.

Real number
Imagine a line marked with an endless series of numbers, with zero in the center, the numbers 1, 2, 3 and so on (forever) written to right of the zero, and -1, -2, -3 and so on written to the left of it. Although fractions like ½ are not written on the line, you could tell just where they are. Irrational numbers like $\sqrt{2}$ have places on the line, too, even though their positions aren't exactly known ($\sqrt{2}$ is somewhere between 1.414 and 1.415). Transcendental numbers, like π, are there, too. This is the number line, and everything on it is a real number. Imaginary numbers (which are based on $i = \sqrt{-1}$) are not on the line.

Theorem
An interesting or useful mathematical statement that has been proved to be true.

Transcendental number
A number which cannot be exactly calculated from an equation. π is a transcendental number: although there are series which can give π to any required accuracy, they are infinitely long and therefore never give an exact answer.

Variable
A letter in an equation which can take a number of different values.

Whole number
A number in the series 0, 1, 2, 3 ...

Index

Symbols

π 32, 39, 41, 43, 47, 73, 88

A

abstract algebra 151, 158–159
acceleration 115
Achilles and the tortoise Paradox 167–168
aeolipile 48
Alexandria, Egypt 35–36, 55, 59–60
algebra, definition of 60
algorithm 61–63
Al-Khwarizmi 6, 60, 62, 64, 66–67
Analytical Engine 155
ancient Egypt 15, 26–27, 30, 35, 55, 67
ancient Greece 18–19, 24, 26, 28, 32, 34–37, 39, 40–42, 46–47, 50, 52, 55, 59, 60, 62, 75, 77, 82–83, 88, 110, 138, 167
antilogarithm 128–129
Arabian mathematics 53
Archimedes 41–45, 56, 64, 75, 89, 133
Argand diagram 81, 135, 148
Aristotle 152–153
arithmetic 6, 62, 72, 77, 106, 144, 150
Arithmetica 47–48, 50, 52–53, 62, 66, 99, 100
artillery 116

B

Babbage, Charles 155
Babylonian mathematics 10–16, 34, 42, 53, 62, 64, 82, 84, 140
Babylonian numbers 11
bacteria 126, 128–129
Baghdad 60, 62

Barber's Paradox 171
base 10 11, 14, 129, 155
base 60 11, 14
Bernoulli, Jacob 124, 126, 133
Big Bang 164
binary 153–155, 160–161
Bombelli, Rafael 80–81
Boole, George 152–155
Brouwer's fixed point theorem 97

C

calculators 102–103, 172
calculus 6–7, 9, 38, 40–43, 45, 55, 86–89, 91, 98–99, 110–113. 115, 118, 129, 136–139, 159, 166, 168
Cardano, Gerolamo (Hieronymus) 77–80, 84
Catalan, Eugène 51
Cauchy, Augustin-Louis 136–140, 144, 159, 168
Cayley tables 145
centroids 56–59
Chain of Pappus 89
China 60, 62
ciphers 85
circles 16–17, 19, 38, 39–41, 43, 45, 55–57, 61, 64, 65, 73, 96, 113, 132, 149
Clay Mathematics Institute 121, 174
coded messages 85
comet 9, 117
complex numbers 52, 80–81, 131, 135, 148–149
contrapositive 6, 22, 23
contrapositive proof 22
convergent series 73
counting numbers 32
cubic equations 42, 44, 64–65, 74–77, 130, 133
cuboid 46, 51
cuneiform 10, 12

D

decimal system 62
Delian problem 39
Descartes, René 92–95, 97, 102, 107, 131
determinate 47
diagonals 28–29, 51, 86–88, 105
diameter 19, 65, 87
differential equations 6, 116–117, 118–119, 121, 128
differentiation 88, 113–115, 138
dimension 45, 54–55, 57, 59
dimensions 37, 40, 51, 55, 86, 119, 135, 148, 150, 175
Diophantine equations 51
Diophantus 47–53, 62, 64, 66, 75, 80, 99, 100, 129, 158
direct proof 18
discriminant 74, 84, 131
divergent series 73
donut 55

E

economists 9, 18
Einstein, Albert 160, 162, 165, 170
electronics 156–157
elliptical equations 100
Empress Catherine the Great 123
engineers 9, 14, 18, 121, 137
equilateral triangle 142, 144–146, 160–161
Eratosthenes 63
Euclid 24–25, 34–36
Euler, Leonhard 51, 122–126, 133, 138
even number 18, 20

F

factorization 6
Fermat, Pierre de 52, 98–102, 158, 173

Fermat's Last Theorem 98–101
Fibonacci 62, 66–70, 107
Fibonacci sequence 68–70
function 47, 87–88, 111, 114, 126, 139, 174

G

Galileo 36–38, 77
Galois, Évariste 140–142, 144, 159
gambling 78, 102
Gauss, Carl Friedrich 70, 72, 130, 133–135
general relativity 160
geometry 6, 24, 34, 36–38, 44, 46, 75, 82–83, 93, 96–97, 142, 166
Germany 10, 123, 160
Gödel, Kurt 157, 166–169, 170–174
graphs 87, 93, 97, 144
Great Library of Alexandria 35
group theory 32, 130, 144–145, 159, 161, 174

H

Hamilton, William Rowan 148–151
Heron of Alexandria 46–48
Hieron 42, 44
Higgs boson 145, 159
Hilbert, David 52–53, 168
Hippasus 30
House of Wisdom, Baghdad 60, 62
hypotenuse 28–31, 43, 149

I

i 79, 81, 85, 94, 114, 131–132, 148–151
identity 37, 47, 161
imaginary number 9, 79–81, 130–131, 150
indeterminate 47
India 22, 60, 62, 67
induction 20–21

infinitesimals 98–99, 166, 168
infinity 24, 73, 160, 168, 174
integers 32
integral 41, 55, 114–115, 117, 126–128
integration 6, 41, 114–115, 117–118, 126, 138
irrational numbers 30, 32, 46, 166
Islamic mathematics 50, 60, 62
isosceles triangle 19, 149

J
Jacquard loom 155–156

K
Kepler, Johannes 86–91, 134, 167
Khayyam, Omar 64–67

L
Lagrange, Joseph-Louis 52
Leibniz, Gottfried 99, 110–111, 115, 118, 137, 153–154
Leonardo of Pisa 66
Liber Abaci 66–67
logarithms 52, 128–129
logic 23, 153–156, 169
Lovelace, Ada, Countess of 155

M
Millennium Problems 174–175
modular forms 101
modulo arithmetic 144
momentum 162–163

N
natural logarithm 128
Navier–Stokes equations 119–121
negative numbers 13, 49, 52, 75, 79, 130
New Solid Geometry of Wine Barrels 90

Newton, Isaac 9, 38, 99, 110–112, 115, 117–118, 133–134, 137, 152, 159, 169
Noether, Emmy 159–160, 162, 165
numerologists 31

O
odd number 18

P
Pappus of Alexandria 55
Pappus's hexagon theorem 58
parabola 41, 65
parallelograms 64
Paris Academy of Sciences 122
Pascal, Blaise 65, 73, 102–109, 111
Pascaline 102–103, 106
Pascal's Triangle 65, 73, 102–103, 105–107, 109
Pascal's Wager 108
perimeter 30, 43
permutations 144, 146
Persia 60
Piazzi, Giuseppe 134–135
pinecones 69
polynomial equations 130, 140
positive numbers 49
prime numbers 24, 51, 63, 174
probability 78, 102
proof by contradiction 23
proof by example 25
proof by induction 20
Pyramids of Giza 15
Pythagoras 12, 15, 26–30, 32, 43, 51, 53, 62, 66, 88, 100, 149
Pythagoras's theorem 12
Pythagoreans 15–16, 26–32, 33, 51, 108
Pythagorean triples 15, 51

Q
quadratic equations 12–14, 32, 74–75, 80, 83–84, 111, 131–133, 142, 159
quark-gluon plasma 164
quartic equations 140
quaternions 148, 150–151, 161
quintic equations 133, 140–142, 144, 159

R
radius 40, 55, 57, 88, 96
rational numbers 27–28, 30, 32, 52, 160
right-angled triangle 26, 29–30, 43, 149
Roman numeral system 71
roots 8, 12, 14, 39, 44, 65, 79, 84, 130–132, 140, 150
rope stretchers 30
Royal Society of London 112
rules of Pythagoras 33
Russell, Bertrand 157, 166–167, 169, 170–173

S
scalar 151
Sieve of Eratosthenes 63
snowflakes, shapes of 90–91
solutions, see roots 44, 49, 51, 53, 74–77, 84, 97, 119, 121, 135, 138, 140, 174–175
spacecraft 114–115
square numbers 12, 37, 52
square roots 12, 14, 74–75, 79, 84, 129, 131, 149
squaring the circle 39
St. Petersburg Academy 123
symbols 6, 10, 12, 36, 38, 48, 71, 80, 82–83, 151, 154, 172

symmetry 54, 55, 101, 142–147, 160–164
Syracuse, Sicily 42, 44

T
Taniyama-Shimura conjecture 100–101
Tartaglia, Niccolò Fontana 74, 76–77, 78, 79
Thales 17–18, 19, 21, 28
Thales' theorem 19
The Laws of Thought 152
The Principia 110, 118
time reversal asymmetry 147
time reversal symmetry 147
torus 55, 90
Tower of Babel 12
transcendental numbers 32
triangle 19, 26, 28–30, 38, 41, 43, 46–47, 54, 57, 104, 106–108, 109, 142–146, 149, 160–161
trigonometry 43
trisecting an angle 39
types of number 32

V
variable 47, 83, 88, 92, 113–115, 117, 125, 138, 168
vector 151
velocity 47, 114–115, 117–118, 151, 163
Viète, François 82–85, 93, 99, 150, 158
volume 34, 39, 41, 46, 55–57, 65, 86–90, 168, 169

W
whole numbers 32
Wiles, Andrew 100–101

Z
Zeno of Elea 68–75, 94, 106, 138, 148–149, 159–160, 166–167

© 2019 Shelter Harbor Press.

All rights reserved. No part of this publication may be reproduced, stored in a retrieval system, or transmitted, in any form or by any means, electronic, mechanical, photocopying, recording, or otherwise, without prior written permission from the publisher.

Cataloging-in-Publication Data has been applied for and may be obtained from the Library of Congress.

ISBN 978-1-62795-117-3

Design: Bradbury & Williams
Copy editor: Meredith MacArdle
Proofreader: Julia Adams
Consultants: Kevin Adams & Alex Teckenbrock
Picture Research: Clare Newman
Cover Design: Wildpixel LTD

Publisher's Note: While every effort has been made to ensure that the information herein is complete and accurate, the publisher and authors make no representations or warranties either expressed or implied of any kind with respect to this book to the reader. Neither the authors nor the publisher shall be liable or responsible for any damage, loss, or expense of any kind arising out of information contained in this book. The thoughts or opinions expressed in this book represent the personal views of the authors and not necessarily those of the publisher. Further, the publisher takes no responsibility for third party websites or their content.

Shelter Harbor Press

603 West 115th Street Suite 163
New York, New York 10025

For sales, please contact
info@shelterharborpress.com

Printed in China.

10 9 8 7 6 5 4 3 2

PICTURE CREDITS

INSIDE: Pg 4-5: All Image repeat use from inside; **Alamy**: AF Fotografie 34, Age Fotostock 169, Artokoloro Quint Lox Ltd 136b, Chronicle 10, 17b, 36, 45, 110cl, 137, Classic Image 126tr, Colport 72, Ian Cook/All Canada Photos 26, Everett Collection Historical 170b, Paul Faern 47, 60t, GB Images 94tr, Interfoto 6c, 22, 53b, 60brb, 123, Sebastian Kaulitzki 128, Lebrecht Music & Arts Photo Library 99, North Wind Picture Archives 48, 116br, Old Paper Studios 78bl, Zev Radovan/Bible Land Pictures 30, Science History Images 7b, 62, 71, 74b, 103b, 125, Alexander Tolstykh 18, Universal Images Group/North America LLC 98bl, World History Archive 112; **Archive.org**: 83, 90; **CERN**: 164; **Clay Mathematics Institute**: 121trb; **Mary Evans Picture Library**: 42, 116tl, 141b, 158; **NASA**: 134tr; Public Domain: 50, 170t; **Science Photo Library**: Max Alexander/Trinity College, Oxford 91cr, Professor Peter Goddard 100br; **Shutterstock**: Nata Alhontess 86bc, Radu Bercan 108b, Darsi 154, Paul Fleet 163, Iryna1 118br, Lenscap Photography 173, Zern Liew 663, Makars 69b, Valemtymc Makepiece 31, Marzolino 98tr, Mattes Images 103t, Militarist 156cr, Morphart Collection 44cr, Oksana2010 55, Rasoulati 14, Roman Samokhin 97cr, Sensay 97br, Roman Sotola 121trt, Torook 64t, Tomer Tu 44tl, Urfin 126b, Natalia Vorontsova 142, vrx 133, Waj 15, Igor Zh 21; **The Wellcome Library, London**: 38, 52, 92tl, 102, 111, 122c, 136cl, 167; **Thinkstock**: Baloncici 91br, Bazilfoto 68, Brand X Pictures 54c, Cronislaw 54cr, Tom Cross 115, Digital Vision 54trb, Dorling Kindersley 148, Eurobanks 54trt, iStock 58, 120b, 171b 172, Lilipom 20, Panimoni 23, Photos.com 24 27, 28, 119, 152t, 165, Pure Stock 176, Sashuk9 54bc, Stocktrek 151, Trasja 54brr, Zoonar 54brl; **Wikipedia**: academo.org 113, 6t, 7t, 9t, 12tr, 12bl, 13, 25, 33, 46, 51, 53t, 57, 59cr, 59b, 60brt, 64b, 65, 66t, 67, 74cl, 76t, 76b, 77, 78bc, 79, 80, 82ct, 82cb, 84, 89bl, 89br, 91t, 92c, 93bl, 93bc, 94tl. 95tr, 95bl, 100tl, 108t, 109, 110tr, 117, 118bl, 122t, 124, 126tc, 130, 134tc, 134b, 138bl, 138br, 140, 141t, 150bl, 150br, 152b, 153, 156tl, 156br, 157, 159, 161, 162t, 162b, 166, 174t, 174b, 175t.

Publisher's note: Every effort has been made to trace copyright holders and seek permission to use illustrative material. The publishers wish to apologize for any inadvertent errors or omissions and would be glad to rectify these in future editions.

Tom Jackson, editor, is a science author who specializes in recasting science, mathematics, and technology into lively historical narratives. After 20 years of writing, Tom has uncovered a wealth of stories that help create new ways to enjoy learning. He studied at the University of Bristol and still lives in that city with his wife and three children.